Vasile Berinde

# Exploring, Investigating and Discovering in Mathematics

Springer Basel AG

Author:
Vasile Berinde
Department of Mathematics and Computer Science
North University of Baia Mare
Str. Victoriei Nr. 76
4800 Baia Mare
Romania
e-mail: vberinde@ubm.ro

2000 Mathematics Subject Classification: 97D50; 97C30, 97U40

A CIP catalogue record for this book is available from the
Library of Congress, Washington D.C., USA

Bibliografische Information Der Deutschen Bibliothek
Die Deutsche Bibliothek verzeichnet diese Publikation in der Deutschen Nationalbibliografie;
detaillierte bibliografische Daten sind im Internet über <http://dnb.ddb.de> abrufbar.

ISBN 978-3-7643-7019-0 ISBN 978-3-0348-7889-0 (eBook)
DOI 10.1007/978-3-0348-7889-0


Originally published by Birkhäuser Verlag, Basel – Boston – Berlin in 2004
Cover design: Micha Lotrovsky, CH-4106 Therwil, Switzerland
Printed on acid-free paper produced from chlorine-free pulp. TCF ∞

ISBN 978-3-7643-7019-0

9 8 7 6 5 4 3 2 1 www.birkhauser-science.com

In the memory of my father

# Preface

We present here the English version of the Romanian first edition (V. Berinde: *Explorare, investigare si descoperire in matematica*, Editura Efemeride, Baia Mare, 2001). There are no major changes. Only a few printing errors were corrected. When transcribing Romanian names or denominations we did not use the diacritical marks.

Our purpose is to provide an introduction to creative problem solving techniques with particular emphasis on how to develop inventive skills in students. We present an array of 24 carefully selected themes that range over all the main chapters in elementary mathematics: arithmetic, algebra, geometry, analysis as well as applied mathematics. Main goal is to offer a systematic illustration of how to organize the natural transition from problem solving activity toward exploring, investigating and discovering new facts and results.

The book is addressed mainly to students, young mathematicians, and teachers, involved or/and actively working in mathematics competitions and training gifted people. It collects many valuable techniques for solving various classes of difficult problems and, simultaneously, offers a comprehensive introduction to creating new problems.

The book should also be of interest to anybody who is in any way connected to mathematics or interested in the creative process and in mathematics as an art.

Among the particular features of the book we mention: fully creatively worked-out source problem(s), for each theme; emphasis on teaching natural exploration and investigation skills, starting from a source problem and its solution, that finally lead readers to discover new interesting and challenging problems, that invite further explorations and so on; the exposition is at an elementary level but it opens the way to both elementary and higher mathematics research work.

All 24 themes included in the book contain new information and original results, presented as much as possible in the framework of their natural creation process. The heuristic approach we used, based on generalization, generality and algorithmicity principles, as parts of the whole creative process, is briefly described in the Addendum. All these themes, as a whole, contribute to shaping the readers' inventive behaviour.

At the first level of reading, the book offers a significant number of accessible methods for solving difficult (sometimes very difficult) problems. The book assumes only a basic knowledge of elementary mathematics and provides worked solutions to over 100 problems, and hints or clear ideas of solving over 150 problems and exercises, more or less related to the first ones.

At the second level of reading, the book aims to develop creative and primary research skills in readers. In this respect, it can fill a serious gap in mathematics education: after many years of predominantly non-creative problem solving activity, graduate students are suddenly expected to do their first research work — a task for which they have not been properly prepared. In the author's hopes, the present book could be an ideal starting point for learning how to perform such research work.

If desired, this book can be used for a lecture course on Methodology of Mathematical Research / Discovering in Mathematics with little or no further preparation.

I should like to thank my daughters, Madalina and Ruxandra, who worked hard to translate the book into English. I am grateful to Tony Gardiner from the University of Birmingham who commented on early sections.

September 2003
Baia Mare

Vasile Berinde

# Contents

# Introduction

Many years have passed since the idea of writing this book took precise shape, and another five or six years were needed to write it, thus the entire process of "making" it covers almost fifteen years. In spite of this, the book is not a very elaborate one. My first intention was to publish it in 1995, the year when the journal *Gazeta Matematica* celebrated a century of existence. But, because of overwhelming administrative duties at that time (duties that consumed considerable time and intellectual energy), I had to interrupt my work for many years. I only began to work again at this book in 2000, *The World Mathematics Year.* I do not consider that the book is finished now as it was initially planned, but this is the best written version to date.

**How did the idea of writing this book evolve?**

It was simply a corollary of my activity as a problem proposer for mathematics journals and contests. Now I have to mention that I only sporadically dealt with this part of elementary mathematics, as an immense nostalgia for the times of my training as a problem solver of *Gazeta Matematica.* An important role in favoring elementary mathematics was also played by *Lucrarile Seminarului de Creativitate Matematica*, a journal that was founded in 1991 exactly with the aim of emphasizing the creative approach in practicing mathematics.

**What Does the Book Deal with?**

As the title suggests, the book is intended to be an introduction to mathematical research by means of some elementary themes. I have been asked many times, by pupils, students and high school teachers, if there is anything left to be discovered in mathematics, which they took for a completely formed field, with no other development possibilities.

Well, this book is a wide and insistent affirmative answer to this question. Its aim is to offer a concrete and as exact as possible description of the act of discovery in mathematics.

Using examples which I consider appropriate and that were also close to my own interests, I tried to show the nature of a creative approach to any mathematics problem. Thus I have tried to capture the attitude, the "mind-set", of a researcher contemplating a problem, whether that problem has been chosent a priori or has arisen in the context of a related question. This seems to me the natural behavior of any researcher, whatever his/her field of activity might be.

When facing a problem (in mathematics, technology, arts, etc.) one has essentially the same attitude as assumed in any creative process. First of all one must know the problem thoroughly, then probe it from all possible angles, looking for its essential features, its deepest meanings, and the reasons for its behavior. Then one must develop conclusions from which to push the evaluation and investigation of these aspects as far as possible towards not only a solution but a connection with the wider theme.

The steps of this creative approach are inherent in the very title of the book: "Exploring, Investigating and Discovering in Mathematics". We first explore the inner structure, the "deposit of precious metals" of a certain problem, then, once the "mother lode" has been found, we investigate all, or as many as possible, suggested directions; if our creative approach has been well conceived, at the end of these efforts the result, the "precious metal", will certainly appear — and that is the discovery.

This same creative approach can be applied to higher mathematics or to any field of scientific research. To demonstrate this point without decreasing the accessibility of the book, some problems in Sections 23 and 24 go beyond the elementary level.

### To Whom is This Book Addressed?

This book was written so that it can be read (at least in some of its parts) by anyone with an interest in and some study of mathematics. But of course it is mainly addressed to those who practise mathematics as a scholarly subject: students, teachers and researchers.

As already mentioned, it is first of all a guide for those who are starting their scientific research and need to develop a precise working method. Reading and thinking about one or another of the themes in the book might well lead a reader's first steps in mathematical investigation, showing how a problem is born and gives birth to other problems — demonstrating how one can always push the level of his/her knowledge, no matter how small and limited the original research theme might have been.

The 24 themes included in this book touch upon almost all elementary mathematics, starting with arithmetic and continuing through algebra, geometry, calculus and even applied mathematics. They reflect not only the authors personal tastes and professional interests but his commitment to elementary mathematics — a tribute to his training as a problem solver with the outstanding monthly *Gazeta Matematica*.

**What is the Structure of the Book?**

The book comprises 24 chapters and an *Addendum* that tries to synthesize the main principles that have guided me in developing my own problem solving method. Each chapter deals with a specific "theme". Each theme is treated with a number of variations, similar to the techniques of musical composition, or perhaps to the action of a spider, weaving a web of reflections in an infinite set of mirrors. Starting with a known source, a chain of variations, or of reflections, creates a new image, often at least as valuable as the one that was the starting point. In science as in art, the creative principles are essentially the same.

In its turn, each theme (chapter) has a more or less fixed structure, depending on the material and on the imagination and inspiration of the investigator. The following arrangement of the material can be found in almost every theme:

1. The text of the source problem (the starting problem)
2. The solution (or solutions) of the source problem, highlighting the creative principles that this solution (solutions) are based on:
   (a) algorithmicity;
   (b) generality (of the method);
   (c) generalization (of the data and requirements);
3. Remarks regarding the essence of:
   (a) the problem;
   (b) the given solution;

   and the outline of the first exploratory directions;
4. The investigation of the directions outlined in 3, considering:
   (a) the generality principle;
   (b) the generalization principle;
5. Finalizing the explorations, systemizing the conclusions of all effectively finalized investigations and outlining the results by:
   (a) reformulating the problem;
   (b) formulating new problems (which are already solved);
   (c) possibly building a "factory" of new problems;
   (d) setting up a list of related problems (which the reader is invited to solve);
   (e) suggesting other possible investigation directions;

6. Bibliographical comments about the source problem, similar or related problems, books or articles where the general theory, on which the problem is based, is grounded, and suggested further reading.

We do not pretend that our references are complete. On the contrary, we restricted them to the Romanian mathematics journals to illustrate how certain themes developed exclusively within our own mathematics literature: obvious treatments of some of the themes can be found in the international literature as well.

**How original is the book? What are its merits?**

The author found a great joy of the spirit in writing this book, the sensation of expressing the very core of his ideas about mathematics and research work. As no research is exclusively the domain of an individual, it is easy to see how original such a book can be. We claim originality for our book mainly through the author's conception of solving problems, a conception that is illustrated through some themes that have been of interest to him for varying periods of time. There are other publications dealing in a similar manner with elementary mathematics problems (the best known of which is G. Polya's book *Discovery in Mathematics*), still we were guided here by our own method, presented in the *Addendum.* How original and useful it is, is left to the reader to decide.

The goal of originality also motivated us to avoid the use of some common terminology, e.g., "the heuristic method", etc. Instead we preferred to use, when possible, concepts that are much closer to the creative act in mathematics as we see it, and that could be summarized as follows: never when we finish writing down the solution of a mathematics problem should we be tempted to state that the problem is now completely solved. No way! "Exhaustive" is a term that every mathematician should avoid. As there is always something new to be said related to a solved problem, new approach angles, new investigation and exploratory directions can appear at any moment.

There is still to be said that a lot of the results and "discoveries" of this book belong equally to other mathematicians, and strictly limiting or deciding upon the priority of one or another was not the intention of the author. Still, in most the cases the presentation itself will suggest the evolution of ideas on that theme.

And finally I wrote this book hoping that it would express one of my fundamental beliefs regarding the human spirit.

I strongly believe in the existence of some latent creative resources in each of us — sparks of our creator and reflections of the divine act of creation. I would even dare to say that these latent creative resources pre-exist in the human being. An adequate methodological framework, and especially the tools specific to creation, are necessary to awaken them. No matter if he is an artist, a poet, a mathematician, an engineer, a physicist, a chemist, a tailor, a cook or a peasant, the one who seeks the way to creation, learns how to use these tools and finds the secret of creation will attain spiritual self-fulfillment.

Succeeding in a profession, whatever that profession might be, depends mainly on finding the tools necessary for revealing this creative part of ourselves. Blessed are those who have come to this realization, obtained the appropriate tools, learned their functions, and take the time to use them intensely for creation!

Vasile Berinde

Donetsk, Ukraine
29th of May – 1st of June 2001

# Theme 1

# Chase problems

This section explores several methods for solving a class of problems called **chase** or **chasing problems** (known also under the name **pursuit problems**), as well as to illustrate the way these explorations and investigations can be creatively carried on in order to get more information on the whole class of these problems and, especially, in order to obtain new problems and results.

Chase problems model concrete situations we can meet in practice: two moving trains on two parallel railways, racing (horse racing, bicycle racing, motor racing etc.), hunting, and so on. Consequently, the statements of these problems could be different from one case to another, but their mathematical essence is the same and is completely described by the following problem, which is the starting point of all investigations given in the present section.

## 1.1 Source problem

**Problem 1.** *A greyhound chases a fox which is* 60 *fox jumps ahead of it. Three greyhound's jumps are as long as seven jumps of the fox, but while the greyhound does* 6 *jumps the fox does* 9 *jumps. Can the greyhound catch up with the fox? How many jumps has the greyhound to run in order to catch up with the fox?*

**Solution.** Let us denote the unknown data of the problem by

$x=$ the number of jumps that the greyhound has to run in order to catch up with the fox,

$y=$ the number of jumps the fox runs from the moment when the greyhound starts the chase until it catches up with the fox.

First of all, let us notice that the data of this problem, which in the sequel will be called **the chasing parameters,** can be grouped into two categories: data concerning the **time component** of the chase, on the one hand, and data concerning the **spatial component** of the chase, on the other hand.

By separating them, we deduce that during a **time** unity

**while** the greyhound does 6 jumps, the fox does 9 jumps.

Therefore, during the whole chase, the number of jumps done by the greyhound $x$ and the number of jumps done by the fox $y$ are proportional to 6 and 9, respectively, that is

$$\frac{x}{6} = \frac{y}{9} \Leftrightarrow 9x = 6y. \tag{1}$$

On the other hand, if we are considering the spatial component of the chase, we know that:

3 jumps of the greyhound are as long as 7 jumps of the fox (see Figure 1).

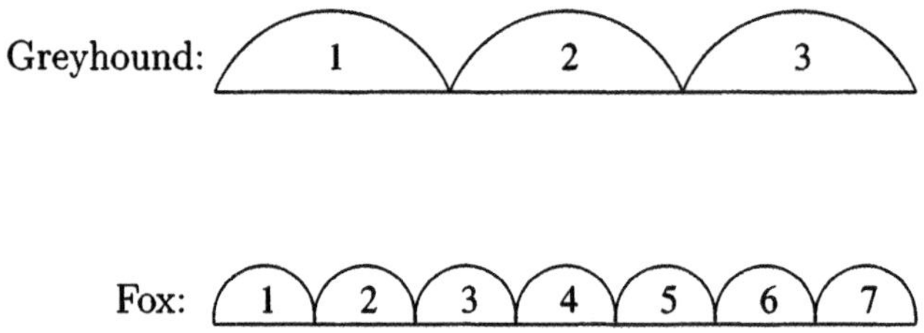


Figure 1

Therefore, **over the same distance,** the number of all their jumps remains in a constant ratio, i.e., 3/7. But, over the whole length of the chase, that is from the point where the greyhound begins the chase till it catches up with the fox, the greyhound does $x$ jumps while the fox does $60 + y$ jumps (it was 60 fox jumps ahead of the greyhound, see Figure 2).

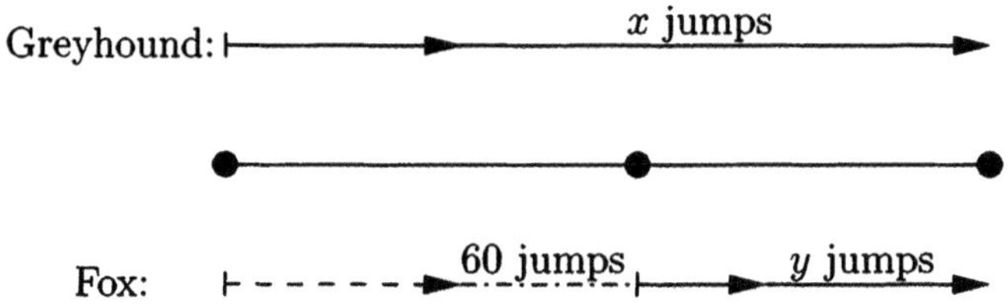


Figure 2

Hence, the ratio $x$ over $60 + y$ is equal to 3/7:

$$\frac{x}{60+y} = \frac{3}{7} \Leftrightarrow 7x = 3(60+y). \tag{2}$$

Therefore, in order to solve Problem 1 we have to solve the linear system (1)+(2):

$$\begin{cases} 3x - 2y = 0 \\ 7x - 3y = 180, \end{cases} \tag{3}$$

which is a compatible system (i.e., the greyhound always catches up with the fox) with the solution

$$x = 72, y = 108.$$

Hence, the greyhound has to run 72 jumps in order to catch up with the fox. □

## 1.2 A Close Examination of the Problem and its Solution

Upon a close examination of the problem data and of the way they were exploited in solving the problem, one notices that

1) All the arguments used in the solution are valid even though the data of the problem are changed. This means that our method satisfies the **generality** condition.

Therefore, we can use the same method for solving a more general problem (e.g., Problem 1a, Section 1.3).

2) The solving method is clearly divided into different steps, that is, the method satisfies the **algorithmicity** condition, which is a typical condition required by any creative solution of a problem.

The main merit of such an algorithmic solution consists of the fact that it may be easily understood by students and then used to solve other related problems.

3) Due to the fact that it is a practical one, Problem 1 allows us, by means of the given solving method, to emphasize the specific steps of any mathematical modelling process. In fact, the most difficult task in solving chasing problems appears exactly at the stage when the physical data of the problem are translated into mathematical representations. By inserting this class of problems in the present book we tried to help the reader discover by himself/herself the three basic steps of any mathematical modelling process:

**M1. Translation of the given physical information** to a mathematical form (algebraic equations, differential or integral equations etc.), that is, the construction of the appropriate mathematical model of the physical situation;

**M2. Treatment of the model** by specific mathematical methods (analytical or numerical methods). This step leads to the solution of the given problem in mathematical form;

**M3. Interpretation of the mathematical results** in terms of the physical phenomenon.

All these three steps seem to be of equal importance in modelling but, in the next section, we will stress the benefits we can get by applying the last one.

As it will be seen, the suggestions given by the physical meaning of the problem will play an important role in the following.

## 1.3 First Direction of Investigation

The remark 1) in the previous section suggests that one possible direction for investigating Problem 1 is the very solution of its generalization. We shall see that, although the generalization of many problems are trivial, that of Problem 1 is very fruitful, due to the additional information on the essence of chasing problems that we can find in this way and, especially, due to several exploring and investigating directions that are suggested.

**Problem 1a.** *A greyhound chases a fox which is N fox jumps ahead of it. While the greyhound does a jumps, the fox does b jumps, but p jumps of the greyhound are as long as q jumps of the fox. Can the greyhound catch up with the fox? In this event, how many jumps has the greyhound to run in order to catch up with the fox?*

**Solution.** Let $x$ and $y$ denote the same quantities as in Problem 1. We deduce that during a **time** unit the greyhound does $a$ jumps, while the fox does $b$ jumps, hence, over the whole length of the chase, we shall have

$$\frac{x}{a} = \frac{y}{b} \Leftrightarrow bx - ay = 0. \tag{4}$$

On the other hand, by considering the **spatial** component of the chase, we know that $p$ jumps of the greyhound equal $q$ jumps of the fox.

Over the whole length of the chase, that is, from the point where the greyhound begins the chase till the point it catches up with the fox, the greyhound does **x** jumps while the fox does **N+y** jumps.

Hence, using the fact that over the same distance the numbers of their jumps remain in the constant ratio $p/q$, we deduce that

$$\frac{x}{N+y} = \frac{p}{q} \Leftrightarrow qx - py = pN. \tag{5}$$

Solving Problem 1a is now equivalent to solving the linear system

$$\begin{cases} bx - ay = 0 \\ qx - py = pN \end{cases}. \tag{6}$$

Taking into account the physical meaning of the quantities $a, b, p, q, N, x$ and $y$, we deduce that all are positive numbers. By one of the known methods (for instance the reduction method) we get from (6) the equation

$$(aq - bp)x = apN. \tag{7}$$

We distinguish now two cases: a) $aq - bp = 0$; b) $aq - bp \neq 0$.

In the first case, equation (7) is equivalent to $0 = apN > 0$, a contradiction. Therefore if

$$\frac{a}{b} = \frac{p}{q},$$

then the system (6) has no solution, which from a physical point of view means that the greyhound never catches up with the fox.

If $aq - bp \neq 0$, from (7) we obtain

$$x = \frac{apN}{aq - bp} \tag{8}$$

and then by (6) we get

$$y = \frac{bpN}{aq - bp}, \tag{9}$$

which is the required solution of the linear system (6).

The quantities $x$ and $y$ given by (8) and (9), respectively, are not always solutions of Problem 1a for, due to their physical meaning, we must have $x > 0, y > 0$ which require the essential additional condition $aq - bp > 0$, equivalent to

$$a/b > p/q. \tag{10}$$

Therefore, if (10) is true, then the greyhound catches up with the fox after running $x = apN/(aq - bp)$ jumps, while if (10) is not satisfied, i.e.

$$a/b \leq p/q, \tag{11}$$

then the greyhound never catches up with the fox. □

## 1.4 Exploring the Investigation Results

$1^0$. A first conclusion that could be drawn from the solution of Problem 1a is that the initial distance between the greyhound and the fox does not have — from a mathematical point of view — any influence on the result of the chase. Only the **chase parameters** are responsible for the success of the chasing: if the ratio of the **time parameters** $(a, b)$ is greater than the ratio of the **spatial parameters** $(p, q)$, each ratio being computed in the sense chasing/chased, then the chase will be successful.

**Example 1.** In the conditions of Problem 1, we have $(a, b) = (6, 9)$ and $(p, q) = (3, 7)$. Since $\frac{6}{9} > \frac{3}{7} \Leftrightarrow 42 > 27$, (10) is satisfied and hence the greyhound catches up with the fox, and this does not depend (theoretically) on the initial distance between them.

For $N = 60$, by (8) we obtain $x = 6 \cdot 3 \cdot 60/(6 \cdot 7 - 9 \cdot 3) = 72$, the solution found in Section 1.1.

Let us consider now an easy modification of Problem 1, given by

**Problem 1b.** *A greyhound chases a fox which is* 2 *fox jumps ahead of it. Knowing that while the greyhound does* 6 *jumps, the fox does* 9 *and that* 5 *greyhound jumps are as long as* 7 *fox jumps, could the greyhound catch up with the fox?*

**Solution.** In this case we have $(a, b) = (6, 9)$ and $(p, q) = (5, 7)$. Since $\frac{6}{9} < \frac{5}{7} \Leftrightarrow 42 < 45$, we conclude that the greyhound never catches up with the fox, despite the very **short** initial distance: only 2 jumps ! □

**$2^0$.** Problem 1 illustrates the case when, even if the greyhound jumps less often than the fox (while the greyhound runs 6 jumps, the fox runs 9), it is still able to catch up with the fox, due to the fact that the length of its jump compensates for that slower speed. But the things behave completely differently in Problem 1b, where 5 greyhound jumps are as long as 7 fox jumps, which cannot compensate for the frequency difference of the two runners.

A small modification is enough in order to ensure the compensation, e.g. replacing 5 by 4, since $\frac{6}{9} > \frac{4}{7} \Leftrightarrow 42 > 36$.

We can imagine that another greyhound, having its jump smaller than the fox jump, could however catch up with the fox if the frequency of its jumps were high enough to compensate for the length difference, as shown by the next example.

**Example 2.** Let us consider the chase of parameters $(a, b) = (3, 2)$ and $(p, q) = (5, 4)$. Then the chase is successful, since $\frac{3}{2} > \frac{5}{4} \Leftrightarrow 12 > 10$.

In this case, the greyhound needs 5 jumps to cover the distance run by the fox in 4 jumps.

**$3^0$.** Let us go back to Remark $1^0$ and try to explain what it means that the initial distance, $N$, between the greyhound and the fox has no influence on the result of the chasing. As the mathematical model of the chase problem is a linear algebraic system, the final result is theoretically determined by the compatibility of this system over the set of real positive numbers, which is governed by condition (10). But only theoretically. Because, from a practical point of view, the result of a chase problem depends also on the physical resources of the two competitors. A chasing situation which requires the greyhound for instance to take 5000 jumps is not realistic. In this way we are led to introduce a new class of chasing problems which will be called **limited chasing problems.** A typical example is given by

**Problem 1c.** *Madalina usually runs* 1200 *strides during her athletics training. She joins Ruxandra, who has already run* 80 *strides when Madalina begins her race. Knowing that* 7 *of Madalina's strides are as long as* 6 *of Ruxandra's strides and that while Madalina runs* 5 *strides, Ruxandra runs* 4, *establish whether Madalina will catch up with Ruxandra during their joint training (one assumes they both start the race at the same point).*

**Solution.** We have $N = 80, (a, b) = (5, 4)$ and $(p, q) = (7, 6)$. As $\frac{5}{4} > \frac{7}{6}$, Madalina could theoretically catch up with Ruxandra, by racing $x = \frac{5 \cdot 7 \cdot 80}{2} = 1400$ (strides). Since her training distance covers only 1200 strides, the answer is in this case negative. □

**$4^0$.** The main benefit of the previous investigations around Problem 1 seems to be the "recipe" for creating new chase problems. Indeed, the relations (8) and (9) on the one hand, and conditions (10) and (11), on the other hand, allow us to build a real "problem factory" of chasing problems.

Therefore, in order to create a chasing problem, we shall proceed to the following steps:

a) We choose the numbers $a, b, p$ and $q$, preferably integers, such that

1) condition (10) holds — for a successful chase problem;

2) condition (11) holds — for an unsuccessful chase problem.

b) We choose the number $N$, the initial distance between hunter and hunted, in such a way that $aq - bp$ divides $N$ (in order to obtain **nice** solutions for the problem, that is, integer numbers).

c) We choose the limits of the chase (only for limited chasing problems).

We invite readers to enjoy their owning of this "chase problem factory" by composing for themselves as many problems as they wish.

**5$^0$**. In the spirit of this book, we shall consider that a given problem has been solved **creatively** only if, as in the case of Problem 1, after investigating several research directions suggested by its statement and solution, we are able to get new deep information on that class of problems and possibly even to discover new problems.

For the moment we get only one kind of new problems, the **limited chase problems,** but as it will be seen in the following this is not the only benefit of our creative approach.

Let us group the data in a chase problem as follows:

$$\textbf{Problem}:\ N,\ (a,b),\ (p,q)\ = x \tag{12}$$

or

$$\textbf{Problem}:\ N,(a,b),\ (p,q) = y, \tag{13}$$

in the case that the unknown is represented by the number of jumps run by the hunted. With this convention, Problem 1 will be briefly written as

$$\textbf{Problem 1}:\ 60,(6,9),(3,7) = x$$

with the solution $x = 72$, while Problem 1b will be

$$\textbf{Problem 1b}:\ 2,(6,9),(5,7) = x$$

with the solution $x = \infty$, and so on.

Now let us notice that all chase problems considered so far contained in the left-hand side of the sign "=" only known quantities, while the only unknown quantity was situated on the right-hand side of "=".

If we denote by $K$ the known quantities and by $U$ the unknown quantity, in both (12) and (13) we have the following distribution:

$$\textbf{Problem}:\ K,(K,K),(K,K) = U. \tag{14}$$

The method we described for solving a chase problem shows that it is possible to replace any $K$ in the left-hand side of (14) by the $U$ from the right-hand side, thus obtaining new classes of chase problems. For instance,

$$\textbf{Problem: } U, (K, K), (K, K) = K \tag{15}$$

corresponds to the situation when we are looking for the initial distance between the greyhound and the fox, knowing the number of jumps needed by the greyhound to catch up with the fox and, of course, knowing the chase parameters.

In this way we shall obtain five new classes of chase problems.

In the following we indicate a few samples of such new problems, which are generally more difficult than the original one. Next we give a problem belonging to the class (15).

**Problem 1d.** *A dog chases a hare. While the dog does* 3 *jumps, the hare does* 5, *but* 4 *dog jumps are as long as* 7 *hare jumps. How many hare jumps was the hare ahead of the dog, knowing that the latter caught up with the hare after* 60 *jumps?*

**Solution.** We have $(a, b) = (3, 5), (p, q) = (4, 7)$ and $x = 60$. Hence, by (8) we get $N = x \cdot (aq - bp)/ap = 5$ (jumps). □

**Problem 1e.** *A greyhound chases a fox. We know that while the greyhound does* 3 *jumps, the fox does* 7 *and that* 2 *greyhound jumps are as long as $n$ jumps of the fox. Find:*

*a) the minimum value of $n$ such that the greyhound catches up with the fox;*

*b) the maximum value of $n$ such that the greyhound never catches up with the fox.*

**Solution.** Even if $N$ and $x$ are not directly involved in the statement of the problem, we easily identify Problem 1e to be in the class: $K, (K, K), (K, U) = K$, that is

$$\textbf{Problem 1e: } N, (3, 7), (2, n) = x$$

(as usual we shall consider $x = \infty$ in the case when the greyhound cannot catch up with the fox). In this case

a) we must have $\frac{3}{7} > \frac{2}{n}$, which is equivalent to $n > \frac{14}{3}$, and hence the required minimum value is $n = 5$, while in the case

b) we must have $\frac{3}{7} \leq \frac{2}{n} \Leftrightarrow n \leq \frac{14}{3}$, which gives $n = 4$. □

**Problem 1f.** *A greyhound chases a fox which is $n$ jumps before it. While the fox does* 3 *jumps, the greyhound does one, but one greyhound jump is four times greater than the fox jump. Find $N$ such that the greyhound catches up with the fox after* 10 *jumps. Is it possible that the greyhound catches up with the fox after only* 5 *jumps?*

**Solution.** We have $\frac{a}{b} = \frac{1}{3}$ and $\frac{p}{q} = \frac{1}{4}$ and so (10) is satisfied.

Hence, no matter the value of $N$, the greyhound catches up with the fox. By (8) we get

$$x = \frac{apN}{a \cdot 4p - 3ap} \Leftrightarrow x = N.$$

Therefore, in the first case $N = 10$, while in the second one $N = 5$. □

**Remark**. It is important to notice directly from the data of Problem 1f that, jumping once, the greyhound comes nearer to the fox with exactly one fox jump. Hence, in order to catch up with the fox, the greyhound has to run the same number of jumps that the fox had in advantage at the beginning of the chase.

**Problem 1g**. *Two worker teams are building up two identical blocks. The second team began working* 30 *days sooner than the first one. Find the number of days needed by the first team to attain the same stage of the work as the second team, knowing that the productivity of the second team is two times better than that of the first one, and that the first team has three times more workers than the second one (besides we suppose that the two teams are qualitatively homogenous).*

**Solution**. We have $N = 30, a/b = 1/2$ and $p/q = 1/3$. As (10) is satisfied, the two teams will reach the same stage of the work in real time. This will happen after

$$x = \frac{apN}{aq - bp} = \frac{apN}{3ap - 2ap} = N \text{ (days)},$$

that is after 30 working days (of the first team). □

We kindly recommend to the readers — no matter if students or teachers — to compose themselves other chase problems belonging to the classes we have already indicated, and to increase their difficulty level by imposing limits to the chasing.

**6⁰**. Now we direct ourselves to another class of chase problems which are formulated, for the sake of simplicity, in terms of athletics races.

**Problem 1h**. *Two runners $A$ and $B$ participate in a marathon race. We know that while $A$ does* 6 *strides, $B$ does* 7 *and that* 5 *strides of $A$ are as long as* 6 *strides of $B$. Find the number of strides the slower runner has run until the moment he is* 100 *strides behind the faster one.*

**Solution**. It is obvious that this last problem does not belong to one of the classes we have already studied, but it can be easily converted to one by stating it in reverse from the end to the beginning of the race.

In this way we have a typical chase problem with the faster runner "chasing" the slower one. The chase will be finished exactly on the race start line.

Therefore, we have a chase problem of type (13) with $N = 10$, $(a, b) = (6, 7)$ and $(p, q) = (5, 7)$ :

**Problem 1h**: $10, (6, 7), (5, 6) = y.$

The solution is then obtained from (9):

$$y = \frac{bpN}{aq - bp} = \frac{7 \cdot 5 \cdot 100}{6 \cdot 6 - 7 \cdot 5} = 3500 \text{ (strides)}.$$

It is obviously due to the inequality

$$\frac{a}{b} > \frac{p}{q} \Leftrightarrow \frac{6}{7} > \frac{5}{6} \Leftrightarrow 36 > 35$$

that the faster runner has run $x = \dfrac{aqN}{aq - bp} = 3000$ (strides) until the moment mentioned in the problem. □

**Remark.** If we modify the statement of Problem 1h in the following way: "Find the number of strides that the slower runner has taken until the moment the faster runner is 100 strides before him", then we have a different problem. Indeed, by reversing the racing direction exactly at the moment required by the problem, in this new chase problem the runner $A$ has to run 100 strides to reach the point where $B$ was at that moment (unlike in Problem 1h, where $B$ was 100 of his strides before $A$ ). Therefore in this case we first have to find how many strides of $B$, say $N$, is $B$ ahead of $A$, knowing that $A$ is 100 strides behind $B$.

But, on the same distance, the ratio of their number of strides is $\dfrac{p}{q}$, i.e., $\dfrac{5}{6}$, and hence $\dfrac{100}{N} = \dfrac{5}{6}$, which gives $N = \dfrac{6 \cdot 100}{5} = 120$.

This means that $A$ will catch up with $B$ after $B$ has run $y = \dfrac{bpN}{aq - bp} = \dfrac{7 \cdot 5 \cdot 120}{1} = 4200$ (strides).

$7^0$. The main idea of the solution of Problem 1h, converting it to a different chase problem, suggests some remarks concerning infinite chasing problems, that is, chase problems for which condition (11) is satisfied.

If in (11) the equality holds, i.e., $\dfrac{a}{b} = \dfrac{p}{q}$, then the initial distance between the two runners will remain the same all the time, so that this case is not interesting at all. But if in (11) the strict inequality holds, i.e., $\dfrac{a}{b} < \dfrac{p}{q}$, then the distance between the two runners increases continuously. It is hence possible to formulate problems of the type 1h, where the requirement is to find the time in which they get to a certain distance from one another.

**Problem 1i.** *Let the chase parameters be the same as in Problem 1b. If they start their race at the same point, after what number of jumps will the fox be 70 jumps ahead of the greyhound?*

**Solution.** Since $\dfrac{6}{9} < \dfrac{5}{7}$, the greyhound cannot catch up with the fox and therefore the distance between them will increase continuously. We reverse the chasing sense by considering that the fox "chases" the greyhound.

The data in this new problem will be:

$$N = 70; (a,b) = (9,6) \text{ and } (p,q) = (7,5).$$

Since $\frac{9}{6} > \frac{7}{5} \Leftrightarrow 45 > 42$, the fox "catches up" with the greyhound after running

$$x = \frac{aqN}{aq - bp} = \frac{9 \cdot 5 \cdot 70}{9 \cdot 5 - 6 \cdot 7} = 1050 \text{ (jumps)},$$

while on this distance the greyhound has run

$$y = \frac{bpN}{aq - bp} = \frac{6 \cdot 7 \cdot 70}{3} = 910 \text{ (jumps)}. \qquad \square$$

$8^0$. The great majority of the chasing problems we find in physics textbooks have a more simple statement: we know the speeds $v_1$ and $v_2$ and the initial distance and have to find the time they reach certain distances from each other as, for instance, in the next problem.

**Problem 1j**. *Two cyclists, having their average speeds $v_1 = 45km/h$ and $v_2 = 40km/h$, respectively, start simultaneously and at the same point. After what time will the distance between them be 10 km?*

**Solution**. Generally, the distance $d$ is attained after

$$d/(v_1 - v_2) \tag{16}$$

time. In this case, the answer is

$$\frac{10km}{45km/h - 40km/h} = 2h(ours). \qquad \square$$

$9^0$. The singularity of the problems of type 1–1i consists mainly in the difficulty of converting the data expressing the frequency and length of jumps in terms of the speeds of the two competitors.

In the following we shall translate the chase parameters $(a,b)$ and $(p,q)$ into the corresponding speeds, $v_1$ and $v_2$, obtaining in this way the relation (8) directly from (17).

Let $l$ be the length of a greyhound jump. Therefore, the length of a fox jump is $\frac{p}{q} \cdot l$ ( for $q \cdot \frac{p}{q} \cdot l = pl$ ).

During a time unit, the greyhound does $a$ jumps, while the fox does $b$ jumps, hence the distance covered by the greyhound (in a time unit) is

$$a \cdot l$$

and by the fox

$$b \cdot \frac{p}{q} \cdot l.$$

Therefore, the speed ratio will equal the ratio of distance covered in a time unit, i.e.,

$$\frac{v_1}{v_2} = \frac{a \cdot l}{b \cdot \frac{p}{q} \cdot l} \Leftrightarrow v_1 = \frac{aq}{bp},$$

hence

$$\frac{v_1 - v_2}{v_2} = \frac{aq - bp}{bq} \Leftrightarrow v_1 - v_2 = \frac{aq - bp}{bp} \cdot v_2. \tag{17}$$

On the distance covered by $N$ fox jumps, the greyhound does $N \cdot \frac{p}{q}$ jumps; hence the distance between the fox and the greyhound, expressed in length unit, is $N \cdot \frac{p}{q} \cdot l$, since $l$ is the length of a greyhound jump.

Using (16), which expresses the chasing time, and (17), we get that the chasing time is

$$t = \frac{d}{v_1 - v_2} = \frac{N \cdot \frac{p}{q} \cdot l}{v_1 - v_2} = \frac{n \cdot \frac{p}{q} \cdot l}{\frac{aq - bp}{bp} \cdot v_2} = \frac{N \cdot \frac{p \cdot l}{q} \cdot bp}{(aq - bp) \cdot v_2}.$$

But the time needed for the greyhound to do one jump is

$$t_1 = \frac{\frac{bp}{qa} \cdot l}{v_2},$$

hence the number of jumps needed by the greyhound to catch up with the fox is

$$x = \frac{t}{t_1} = \frac{apN}{aq - bp},$$

which coincides with (8). □

## 1.5 Bibliographical Comments

Problem 1, which was the starting point of all explorations and investigations in this section, is taken from [5], p. 22, problem 25.XI. A similar problem, but with a simpler statement, can be found in [4] as problem 32 at the Column "Supplementary Problems", p. 131.

We did not succeed in finding the original source of these problems in the Romanian literature, but many other related problems could be found in several collections of problems ([6], [7]), one of them being adapted from *Gazeta Matematica*, no. 11/1982 (problem E:7797, proposed by Gh. Foghis).

The results contained in this section have been presented, in different forms, at various national and international meetings on didactics of mathematics, or have been published in mathematical journals, see the reference list for the most important of them.

## 1.6 References


1. **V. Berinde,** *Two Strategies for Solving Chase Problems.* (Romanian), Lucr. Semin. Creativ. Mat., **3** (1993–1994), 29–42

2. **V. Berinde,** *A Method for Solving Chase Problems.* (Romanian), Gazeta Matematica (Serie metodica), **14 (93)** (1990), no. 3, 168–174

3. **V. Berinde,** *On Some Strategies for Solving Chase Problems.* (Romanian), Foae Matematica, (1997), no. 5, 2–7

4. **C.P. Popovici, I.C. Ligor, V. Alexianu,** *Mathematics. Arithmetic. Algebra. Textbook for VI Form.* (Romanian), Editura Didactică si Pedagogica, Bucuresti, 1992

5. **I. Craciunel, et al.,** *Mathematics. Algebra. Textbook for VIII Form.* (Romanian), Editura Didactica si Pedagogica, Bucuresti, 1993

6. **P. Telinoiu,** *Exercises and Problems of Mathematics.* (Romanian), Editura Porto-Franco, Galati, 1991

7. **I. Olivotto,** *Arithmetic Exercises and Problems for V-VI Forms.* (Romanian), Editura Didactica si Pedagogica, Bucuresti, 1976

8. **V. Berinde,** *A Collection of Creatively Solved Problems which Can Develop the Inventive Skills of Students,* Internat. Conf. on Technology in Mathematics Teaching, Napier University, Edinburgh, 4–7 September 1995

9. **V. Berinde,** *Active Problem Based Learning for Developing the Inventive Abilities of Engineering Students,* Proceed. of the SEFI Seminar, Univ. Twente, 1–3 November 1995

10. **V. Berinde,** *Creative Mathematics,* Final Round of the Flemish Mathematical Olympiad, Vrije Universiteit Brussel, 24 April 1996

11. **V. Berinde,** *A Natural Way to Initiate Students in Mathematics Research: Creative Problems Sessions,* Internat. Congress on Mathematics Education (ICME-8), Univ. of Sevilla, 14–21 July 1996

12. **V. Berinde,** *Research on Developing Creative Skills in Mathematics Education and Teacher Training,* CIEAEM 50, Univ. of Neuchatel, 2–7 August 1998

13. **V. Berinde,** *Exploring, Investigating and Discovering in Elementary Mathematics.* (Romanian), Annual Conf. of Romanian Math. Soc., Univ. of Bucharest, 29 May–1 June 1997

14. **V. Berinde,** *Exploring Around Some Algebra Problems.* (Romanian), Symposium "Modernizing the Mathematical Education in Romania", Abrud, 8–9 May 1999

# Theme 2

# Sequences of Integers Simultaneously Prime

In this section we design a universal algorithm for solving various problems included in the class mentioned in the title of the section. Then we shall use it in exploring the set of solutions for one special problem of this class. At least two reasons are behind our approach:

1) Many problems of this kind have been proposed in recent years at the Romanian Mathematical Olympiad as well as in *Gazeta Matematica*. They were generally proposed to students in forms V and VI and sometimes had a high difficulty level, despite their very simple statements;

2) A problem of this class does not always possess a unique solution. Moreover, by applying the algorithm we use here we want to stress the fact that by solving such a problem we mean not only to find a solution, but **especially** to prove if it is unique or not.

## 2.1 Source Problems

**Problem 2.** *Find* $n \in \mathbb{N}$ *such that* $n, n+4, n+8$ *are all prime.*

**Problem 2a.** *Find* $p \in \mathbb{N}$ *such that the numbers are all prime:*

$$p,\ p^2+2,\ p^2+4,\ p^2+20.$$

**Problem 2b.** *Find* $n, p \in \mathbb{N}$ *such that*

$$n, n+2^p, n+2^{p+1}, n+2^{p+2}$$

*are simultaneously prime.*

**Problem 2c.** *Find the prime number p such that*

$$2p+1, 3p+2, 4p+3 \text{ and } 6p+1$$

*are simultaneously prime.*

**Problem 2d.** *Find all prime numbers n such that*

$$n+2, n+6, n+8 \text{ and } n+14$$

*are also prime.*

**Problem 2e.** *Find the positive integers n for which*

$$n+1, n^2+3, n^3+5, n^4+7, n^5+9$$

*are simultaneously prime.*

**Problem 2f.** *Find the positive integers n for which*

$$n+1, n+3, n+7, n+9, n+15$$

*are simultaneously prime.*

**Problem 2g.** *Find $n \in \mathbb{N}$ such that*

$$n, n+2, n+6, n+14, n+18$$

*are simultaneously prime.*

**Problem 2h.** *Find the positive integers p and n such that*

$$p,\ p+2^n,\ p+2^{n+1},\ p+2^{n+2},\ p+2^{n+3}$$

*are all prime numbers.*

**Problem 2i.** *Find the positive integers n such that*

$$n-12, n-10, n-4, n-2, n+2, n+4$$

*are simultaneously prime.*

**Problem 2i′.** *Find the positive integers n such that*

$$n, n+4, n+10, n+12, n+16, n+22$$

*are simultaneously prime.*

## 2.2 The Solution Algorithm

**Solution of Problem 2.** For $n+4$ and $n+8$ to be prime numbers, $n$ must be odd and $n > 1$ (1 is not a prime number).

Let then $n = 2k+1$, $k \in \mathbb{N}$. The numbers in the problem will be

$$2k+1, 2k+5, 2k+9, \quad k \in \mathbb{N}^*.$$

Let us notice that for $k = 1$ we obtain the solution 3, 7, 11. We try to check whether this is **unique** or not. Therefore we will show that for the rest of the values $k$ may take, at least one number in the given list will not be prime.

We shall also exploit the fact that the smallest number in the solution found above is 3. Depending on this number, 3, any positive integer $k \in \mathbb{N}$ can be written in one of the forms

$$3p, 3p+1, 3p+2 \qquad (p \in \mathbb{N})$$

as the remainder is 0, 1 or 2 when dividing $k$ by 3.

If $k = 3p+1$, then $2k+1 = 3(2p+1)$ is not prime (excepting for the case $p = 0$, when $k = 1$, which gives the already found solution).

If $k = 3p+2$, then $2k+5 = 3(2p+3)$, which is not prime, $\forall p \in \mathbb{N}$, while for $k = 3p$ we find $2k+9 = 3(2p+3)$, which is not prime, $\forall p \in \mathbb{N}$.

Only for $k = 1$, that is $n = 3$, are the three given numbers simultaneously prime. □

**Remark**. Problems 2c, 2d, 2f, 2g and 2i are all of the same kind as Problem 2, while Problems 2b, 2h and 2e are a little bit different from the previous ones. This is the reason for solving them separately.

**Solution of Problem 2c.** The sequence of prime numbers considered in the problem is in fact

$$p, 2p+1, 3p+2, 4p+3, 6p+1.$$

We cannot have $p = 2$ (since $3p+2$ is not prime) or $p = 3$ (since $4p+3$ is not prime). For $p = 5$ we obtain the numbers

$$5, 11, 17, 23, 31$$

which fulfill the requirement. We discuss then the cases suggested by the first component of the solution, that is 5. We could have

$$p \in \{5k, 5k+1, 5k+2, 5k+3, 5k+4 | k \in \mathbb{N}\}.$$

1) If $p = 5k$ and $k > 1$, then $p$ is a composite number. When $k = 1$, we find the solution given above;

2) If $p = 5k+1$, then

$$3p+2 = 3(5k+1)+2 = 15k+5 = 5(3k+1),$$

which is a composite number for $k \neq 0$. For $k = 0$, we obtain the same mentioned solution;

3) If $p = 5k + 2$, then

$$2p + 1 = 2(5k + 2) + 1 = 10k + 5 = 5(2k + 1),$$

which is also a composite number for any $k \neq 0$. For $k = 0$, similarly to the previous case;

4) If $p = 5k + 3$, then

$$4p + 3 = 4(5k + 3) + 3 = 20k + 15 = 5(4k + 3),$$

a composite number, for all $k \in \mathbb{N}$;

5) If $p = 5k + 4$, then

$$6p + 1 = 6(5k + 4) + 1 = 30k + 25 = 5(6k + 5),$$

again a composite number, for all $k \in \mathbb{N}$.

Therefore, the unique value of $p$ that corresponds to the problem is $p = 5$. □

**Be careful !** By applying the same algorithm to Problem 2a, we shall be immediately led, as it will be seen, to a little discovery.

**Solution of Problem 2a.** We need to find all integers $p$ such that

$$p, p^2 + 2, p^2 + 4, p^2 + 20$$

are all prime numbers.

The values $p \in \{1, 2\}$ are not accepted. For $p = 3$ we get a solution. As 3 is the first (smallest) number in the solution sequence, we shall consider

$$p \in \{3k, 3k + 1, 3k + 2 / k \in \mathbb{N}\}.$$

1) If $p = 3k$ and $p > 1$, then $p$ is composite and, if $k = 1$, we find again the previous solution;

2) If $p = 3k + 1$, then $p^2 + 2 = (3k + 1)^2 + 2 = 3(3k^2 + 2k + 1)$, which is a composite number for all $k \neq 0$;

3) If $p = 3k + 2$, then $p^2 + 20 = (3k + 2)^2 + 20 = 3(3k^2 + 4k + 8)$, which is a composite number, for all $k \in \mathbb{N}$. This shows that the unique value for $p$ is 3.

**Remark.** Although the solution of Problem 2a is completely natural for this kind of problems, we wrote it down in detail in order to stress a conclusion that could be drawn from it. Note that, by solving all the Problems 2, 2b, 2c, 2d, 2h and 2i, we exploit the fact that each number in the given sequence becomes a composite number, at least in one of the cases we are discussing. Apart from this rule, when solving Problem 2a we found one number that was not used at all in this filter, and that number is $p^2 + 4$.

What does it mean? It means that, in fact, our problem can be reformulated, to produce a new one. In this way, we reach a first **discovery** of a new problem.

**Problem 2j.** *Prove that the positive integers* $p, p^2+2, p^2+20$ *are simultaneously prime if and only if the numbers*

$$p, p^2+2, p^2+4, p^2+20$$

*are simultaneously prime.*

**Remark.** By exploiting this idea it is clear that, starting from Problems 2, 2a, 2c, 2d and 2h which are all known to possess unique solutions, we can obtain new problems of the same kind as Problem 2j. In the following we give a single sample and invite the reader to formulate — and then to solve — the rest of them.

**Problem 2k.** *Show that the positive integers*

$$p, 2p+1, 3p+2, 4p+3, 5p+4, 6p+1$$

*are simultaneously prime if and only if the numbers*

$$p,\ 2p+1,\ 3p+2,\ 4p+3 \text{ and } 6p+1$$

*are all primes.*

A similar situation is encountered in Problem 2e, where the following numbers

$$n+1, n^2+3, n^3+5, n^4+7, n^5+9$$

are involved.

We notice that for $n=2$ one obtains the solution 3, 7, 13, 23, 41. As usual, the first number in the solution sequence suggests us to analyze the cases

$$n \in \{3k, 3k+1, 3k+2 | k \in \mathbb{N}\}.$$

1) For $n = 3k$, the numbers $n^2+3$ and $n^5+9$ are obviously composite numbers, for all $k \in \mathbb{N}^*(k=0$ is not accepted);

2) For $n = 3k+1$, we have

$$n^3+5 = (3k+1)^3+5 = M3+6$$

which is a composite number, for all $k \in \mathbb{N}$;

3) For $n = 3k+2$ we get the composite number $n+1 = 3(k+1)$, for each $k \neq 0$ (the value $k=0$, that is $n=2$, leads us to the already found solution).

Therefore the unique appropriate value is $n=2$.

**Remark.** We cannot turn a blind eye to the fact that there is a useless number here, too, $n^4+7$. What does it mean? It means that we discovered an improvement of the initial formulation of Problem 2e, in short, we get a new problem.

**Problem 2l.** *If* $n \in \mathbb{N}$ *such that the numbers*

$$n+1, n^2+3, n^3+5 \text{ and } n^5+9$$

*are all prime, then the numbers*

$$n^4 + 7 \text{ and } n^6 + 7$$

*are prime, too.*

**Solution of Problem 2d.** We cannot accept the values $n = 1$ and $n = 2$.

For $n = 3$ we get the sequence 3, 5, 9, 11, 17 which contains the composite number 9.

For $n = 5$ we obtain the sequence 5, 7, 11, 13, 19 containing only prime numbers. We then try to see whether this solution is unique or not. To this end, by considering (remember that 5 is the smallest number in the solution sequence!)

$$n \in \{5k, 5k+1, 5k+2, 5k+3, 5k+4 | k \in \mathbb{N}\}$$

we have to cover the following situations:

1) $n = 5k$, when the first number in the sequence is composite, for each $k > 1$ (for $k = 1$, we obtain $n = 5$, again the previous solution);

2) $n = 5k + 1$, when $n + 14 = 5(k + 3)$ is a composite number, $\forall k \in \mathbb{N}$;

3) $n = 5k + 2$, when $n + 8 = 5(k + 2)$ is composite, $\forall k \in \mathbb{N}$;

4) $n = 5k + 3$, when $n + 2 = 5(k + 1)$ is a composite number $\forall k \in \mathbb{N}^*$ (for $k = 0$ we get $n = 3$, already discussed);

5) $n = 5k + 4$, when $n + 6 = 5(k + 2)$ is composite for all $k \in \mathbb{N}$.

Therefore $n = 5$ is the unique number satisfying the required conditions in Problem 2d. □

**Remark.** For the type of problems considered in this section, the solving method is intimately based on **exploring** and **investigating**, very appropriate terms for the title of our book, we think. We shall see what valuable information we can get in this way!

**Solution of Problem 2i′.** We rewrite the sequence of numbers by denoting $n - 12$ by $n$. We shall have then the numbers

$$n, n+4, n+10, n+12, n+16, n+22,$$

which are more convenient for computations.

Obviously, $n = 1$ cannot be accepted, while for $n = 3$ and $n = 5$ we get composite numbers. If we take $n = 7$, then we find the sequence of prime numbers: $7, 11, 19, 23$ and $29$.

Now we begin investigating to see whether other solutions exist or not.

As 7 is the smallest number in the sequence, we shall consider

$$n \in \{7k, 7k+1, 7k+2, 7k+3, 7k+4, 7k+5, 7k+6 \,| k \in \mathbb{N}\},$$

and shall analyze the following cases:

a) $n = 7k$ $(k \geq 1)$, when $n$ is a composite number for $k > 1$ (for $k = 1$ we find again the solution above);

b) $n = 7k + 2$, when $n + 12 = 7(k + 2)$ is composite $\forall k \in \mathbb{N}$;
c) $n = 7k + 3$, when $n + 4 = 7(k + 1)$ is composite $\forall k \in \mathbb{N}^*$;
d) $n = 7k + 4$, when $n + 10 = 7(k + 2)$ is composite $\forall k \in \mathbb{N}$;
e) $n = 7k + 5$, when $n + 16 = 7(k + 3)$ is composite $\forall k \in \mathbb{N}$;
f) $n = 7k + 6$, when $n + 22 = 7(k + 4)$ is composite $\forall k \in \mathbb{N}$.
In the remaining case, $n = 7k + 1$, the numbers in the problem are:

$$7k + 1, 7k + 5, 7k + 11, 7k + 13, 7k + 17, 7k + 23$$

and no information is available to indicate whether one of them is composite or not. We have to continue our investigations hoping that, as for all previous problems, we will succeed in proving that there is no other solution.

Therefore we analyze patiently the cases

$$k \in \{5l, 5l + 1, 5l + 2, 5l + 3, 5l + 4 \mid l \in \mathbb{N}^*\}$$

(5 is the leading number because $7k + 5$ was one of the previous cases).

We find again one case, $k = 5l + 3$, when it is not possible to decide whether there is a composite number in our sequence or not. By replacing $k = 5l + 3$ we obtain the new sequence

$$35l + 22, 35l + 26, 35l + 32, 35l + 34, 35l + 40, 35l + 46.$$

The next filter is suggested by $22 = 2 \cdot 11$ and therefore we shall consider

$$l \in \{11m, 11m + 1, \ldots, 11m + 10 \mid m \in \mathbb{N}^*\}.$$

Just at this stage, for $l = 11m + 1$ and $m = 4$ we find a sequence which solves our problem:

$$1597, 1601, 1607, 1613, 1619,$$

since it contains only prime numbers.

Therefore, in the case of Problem 2i$'$, there exist more than one solution! □

**Remark.** As it can be noticed, it was not an easy task to obtain the second solution of Problem 21. However, the problem has been proposed several times at various mathematics competitions, form VI, see the bibliographical comments, without having been completely solved in any of the problems books where it has been published.

By these considerations we tried to stress the fact that the solution of any problem must always be brought to an end. Even if its statement is similar to other related problems we could have surprises, as it happened with Problem 2i$'$. By a complete approach which explores as deeply as possible the sequence of that problem, we can get:

1) very important information on the problem itself;

2) new ways of creatively developing such problems, that will be finalized in section 2.4.

The particular difficulty of Problem 2i′ lies in the fact that in the sequences obtained for $n = 127$, $\{127, 131, 137, 139, 143, 149\}$ and for $n = 337$, $\{337, 341, 347, 349, 353, 359\}$, for instance, only one number is not prime: $143 = 11 \cdot 13$ and $341 = 11 \cdot 31$, respectively, and this could be detected only by means of the filter 11.

## 2.3 The Particularity of Problems 2b and 2h

**Solution of Problem 2b.** The prime number $n$ must be odd. Let us consider the first case, $n = 3$, which suggests the set

$$n \in \{3k, 3k+1, 3k+2 \mid k \in \mathbb{N}\}.$$

Hence, we have to study the relationship existing between $2^p, 2^{p+1}, 2^{p+2}$, on the one hand, and the number 3, on the other hand.

For $p = 2l$, the results are $2^p = 2^{2l} = 4^l = (3+1)^l = M3+1$, $2^{p+1} = 2^p \cdot 2 = M3+2$ and $2^{p+2} = 2^{p+1} \cdot 2 = M3+4 = M3+1$, while for $p = 2l+1$ we have $2^p = 4^l \cdot 2 = (M3+1) \cdot 2 = M3+2$, $2^{p+1} = 2^p \cdot 2 = (M3+2) \cdot 2 = M3+1$ and $2^{p+2} = 2^{p+1} \cdot 2 = M3+2$, which show that for any $p \in \mathbb{N}$ at least one of the numbers $2^p, 2^{p+1}, 2^{p+2}$ has the form $M3+1$ and at least one has the form $M3+2$. Therefore, in our sequence

$$n, n+2^p, n+2^{p+1}, n+2^{p+2}$$

we may have only prime numbers just for $n = 3k$ (and $k = 1$), otherwise at least one number would be composite. For $n = 3$, the sequence is

$$3, 3+2^p, 3+2^{p+1}, 3+2^{p+2}.$$

We discuss then the cases

$$p \in \{3m, 3m+1, 3m+2 \mid m \in \mathbb{N}\}.$$

1) For $p = 3m$ we get

$$\begin{aligned} 2^p &= (2^3)^m = (7+1)^m = M7+1, \\ 2^{p+1} &= 2^p \cdot 2 = (M7+1) \cdot 2 = M7+2, \end{aligned}$$

$$2^{p+2} = 2^p \cdot 4 = M7+4$$

and then the number

$$3 + 2^{p+2} = 3 + M7 + 4 = M7$$

is composite, for $m \geq 1$ (in the case $m = 0$, $3 + 2^p = 4$ is composite);

2) For $p = 3m+1$, we have $3 + 2^{p+1} = 3 + 2^{3m+2} = M7$, which is composite, for all $m \geq 1$. For $m = 0$, that is $p = 1$, we obtain the sequence of prime numbers $\{3, 5, 7, 11\}$.

3) For $p = 3m + 2$ we get, as in the previous case, $3 + 2^p = 3 + 2^{3n+2} = M7$ which is a composite number, for all $m \geq 1$.

For $m = 0$, i.c., $p = 2$, we get the sequence of prime numbers $\{3, 7, 11, 19\}$ which gives the second solution — and the last one — of the problem.

Therefore the answer to Problem 2b is $(n,p) \in \{(3,1),(3,2)\}$. □

**Solution of Problem 2h.** Using arguments similar to those in the previous solution, it necessarily results that $p = 3$ and hence the sequence under study will be now

$$3, 3 + 2^n, 3 + 2^{n+1}, 3 + 2^{n+2}, 3 + 2^{n+3}.$$

We shall discuss, as in the previous problem, the cases

$$n \in \{3k, 3k+1, 3k+2 \mid k \in \mathbb{N}\}$$

and we shall get that, except for the case $n = 1$, we always have at least one composite number in the sequence.

Therefore $n = 1$ and $p = 3$ are the required values.

## 2.4 Conclusions

In this section we briefly illustrated the benefits we can get by means of creative solutions of a problem: we may discover new information on the entities involved in it and, moreover, we may discover new problems.

However, if we take a conclusive look at the problems included in this section (except for Problem 2b, of course), we can notice that Problem 2i′ is particularly special. This problem has no unique solution and it is even possible to find an infinite number of solutions — we leave this task to the reader, who could use even the computer for that (see section 21 in this book, for instance).

Before finishing the creative approach to this class of problems, we want to "repair" the "deficiency" of Problem 2i, which asked us to find the numbers $n$ for which

$$n, n+4, n+10, n+12, n+16 \text{ and } n+22$$

are all prime numbers. In order to ensure a unique solution, the idea is very simple: we add one term more. Thus we obtain a new problem that has, this time, a unique solution (we invite the reader to check it).

**Problem 2m.** *Find the positive integers n such that*

$$n, n+4, n+6, n+10, n+12, n+16, n+22$$

*are simultaneously prime.*

We also invite the reader, for training, to tackle the following 10 sequences of integers:

1) $n, n+2, n+4, n+8, n+10, n+14$;

2) $n, n+4, n+8, n+10, n+14, n+20$;
3) $n, n+2, n+8, n+10, n+14, n+20$;
4) $n, n+2, n+4, n+10, n+14, n+20$;
5) $n, n+2, n+4, n+8, n+14, n+20$;
6) $n, n+2, n+6, n+8, n+12, n+18$;
7) $n, n+2, n+6, n+8, n+12, n+14, n+18$
8) $n, n+2^p, n+2^q, n+2^r$, $n, p, q, r \in \mathbb{N}$ and the numbers $p, q$ and $r$ do not have the same remainder when being divided by 3;
9) $\left[\sqrt{n^2+1}\right], \left[\sqrt{n^2+2}\right]+2, \left[\sqrt{n^2+4}\right]+4, \left[\sqrt{n^2+n}\right]+8$, where by $[a]$ we denoted the integer part of the real number $a$;
10*) $n, n+3^p, n+3^{p+1}, n+3^{p+2}$, $\quad n, p \in \mathbb{N}$.

At the end of this section we would like to assure the readers that, by presenting a plentitude of problems of the same kind, we did not intend to exasperate them, but to stress the merits of creatively solving problems, such that as many as possible of their latent possibilities are explored.

## 2.5 Bibliographical Comments

Problem 2 is the first part of problem E:10240* in *Gazeta Matematica* no. 6/1991; Problem 2b is just problem E:10609 in *Gazeta Matematica* no. 7–8/1993, proposed by T. Zvonaru, while Problem 2a was proposed at the Intercounty Mathematics Competition "Tr. Lalescu", Arad, 1987. The other problems in this section are taken from the following sources:

Problem 2c: E:9638*, *Gazeta Matematica* no. 2/1989;
Problem 2d: E:11087, *Gazeta Matematica* no. 12/1995;
Problem 2e: E:10569 *Gazeta Matematica* no. 4/1993;
Problem 2f: Problem V.176, [2];
Problem 2g: Problem V.187, [2];
Problem 2h: E:11167, *Gazeta Matematica* no. 3/1996;
Problem 2i: E:9569*, *Gazeta Matematica* no. 11–12/1988, author D. Codin;
Problem 2m: Problem I, 4.92, [5].

Related to problem 2i′ we presented a communication ( *The Competition Problems Between Natural and Exaggeration*) at the symposium "Days of the National College Vasile Lucaciu Baia Mare", 3rd edition, January 25, 1997.

## 2.6 References

1. **M. Chirciu, et al.,** *Mathematics in School Competitions. 1996.* (Romanian), Editura Paralela 45, 1997

2. **C. Harabor et al.,** *Mathematics. Forms V–VIII. County, Intercounty and National Olympiads.* (Romanian), Editura Scorpion, 1994

3. **A. Herteg, R. Pop, St. Sabau,** *Mathematics Problems Proposed at Competitions in 1990. Form VI.* (Romanian), Baia Mare, 1990

4. —, Collection of *Gazeta Matematica* (1985–1997)

5. **Gh. Drugan et. al.,** *Mathematics in School Competitions for Form V–VIII. Algebra.* (Romanian), Editura Teora, 1998

# Theme 3

# A Geometric Construction Using Ruler and Compass

Problems of geometric constructions using ruler and compass, or only ruler, form a very special class of problems which, in order to be solved, require not only a very good knowledge of basic results in geometry but also special skills and cleverness. Regretably in the last decades geometry has lost the important position that it had in the curriculum of both secondary and high schools in Romania. This is unfortunate because, in order to be well assimilated, geometry requires the student to extensively exercise problem solving skills and invest time in understanding the fundamental truths that lie subtly hidden behind a geometric figure.

Much more than other chapters of elementary mathematics, geometry stimulates and develops the creative and inventive skills of students, see [8], for instance.

It is the aim of this section to illustrate these facts, starting from a problem in the textbook [5], that suggested to us to begin this section with the following source problem.

## 3.1 Source Problem

**Problem 3.** *Construct (using ruler and compass) a triangle having the same area as a given convex quadrilateral.*

**Solution.** Let $ABCD$ be the given quadrilateral, see Figure 3.

**Step 1.** We consider one of the two diagonals, say $AC$.

**Step 2.** Through one vertex of the quadrilateral that does not belong to $AC$, say $B$, we draw a parallel $d$ to $AC$.

Since $ABCD$ is convex, $d$ will intersect both $AD$ and $DC$.

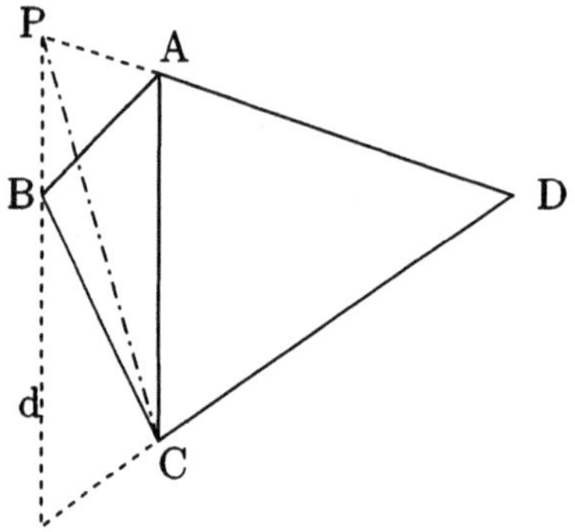


Figure 3

**Important property.** For any point $M \in d$, we have

$$Area\ [MAC] = Area\ [BAC]. \tag{1}$$

**Step 3.** The property (1) immediately suggests how to make the construction. Let $P$ be the intersection point of $d$ and $AD$, $\{P\} = d \cap AD$. Then

$$Area[PAC] = Area[BAC],$$

hence, from

$$Area[ABCD] = Area[ACD] + Area[BAC]$$

it results that

$$Area[ABCD] = Area[ACD] + Area[PAC] = Area[PCD].$$

Therefore $PCD$ is the required triangle in the problem. □

**Remark**. It is well known that all operations needed to solve Problem 3, i.e.:

1) to construct a parallel to a given line;

2) to draw a line passing through two given points,

can be done using ruler and compass.

## 3.2 A Close Examination of the Problem and its Solution

If one carefully reads the data of Problem 3 and the way they were used in the solution, following all steps, one can easily notice that:

1) not all of the arguments used are particularly related to the polygon involved in Problem 3 (a quadrilateral). Only its **convexity** was essential;

2) the solution method itself — being an algorithmic one — satisfies also the generality condition, i.e., it can be applied to another problem similar to Problem 3, that we have just **discovered.**

**Problem 3a.** *Construct a quadrilateral having the same area as a given convex pentagon ABCDE.*

**Solution.** Similarly to Problem 3, let us consider a diagonal, say $BE$. Through the point $A$ (which together with $BE$ forms a triangle cut from the pentagon) we draw a parallel $d$ to $BE$, which will necessarily intersect $BC$ and $DE$, since $ABCDE$ is convex. Let $\{P\} = d \cap BC$. Since

$$Area[ABE] = Area[PBE],$$

the quadrilateral $PCDE$ is the required one. □

**Remark.** By combining Problems 3 and 3a, we have just discovered a new problem, which can be immediately solved.

**Problem 3b.** *Construct using ruler and compass a triangle having the same area as a given convex pentagon.*

Now it is very clear that, following the iterative process indicated for Problem 3a, a hexagon can be reduced to a pentagon of the same area, then the pentagon to a quadrilateral, and so on. This means we are able to solve the following (new!) problem.

**Problem 3c.** *Let $A_1A_2...A_n$ be a given convex polygon. Construct, using ruler and compass, a triangle having the same area as the given polygon.*

**Remarks. 1)** Readers can now compose by themselves a lot of problems of the same kind: one for the transformation hexagon → quadrilateral, another one for octagon → pentagon, and so on, all of them being suggested by the creative solution of Problem 3.

**2)** If we don't find Problem 3c very amusing, for $n$ large enough, say $n = 15$, we can write a computer program to perform these boring operations of successive eliminations, side by side, until we get a triangle. The interested reader may find such a program, written in C language, in the paper [4]. One can find there drawings of successive operations that can be performed by the computer.

## 3.3 Other Directions

We could already be satisfied with the "capture" of the new problems derived from solving Problem 3, but its rich resources have not been exhausted yet. Another direction of investigation is obtained by passing from the plane to three dimensional space.

Thus we obtain the following new problems (we only indicate two of the analogous problems of those previously formulated or suggested).

**Problem 3d.** *Construct a tetrahedron having the same volume as a given prism.*

**Problem 3e.** *Construct a tetrahedron having the same volume as a given convex polyhedron.*

**Hint.** It is quite obvious that, in order to solve these problems, we shall consider a diagonal **plane** and then we shall construct a **plane** parallel to the first one, containing a vertex of the given polyhedron. □

Let us now come back to the plane setting and notice that, by a careful reading of the solution of Problem 3, we discover another problem that can be similarly solved.

**Problem 3f.** *Let $ABCD$ be a convex quadrilateral and $P$ be a point belonging to the prolongation of side $AD$, that is $A \in [PD]$, such that $\widehat{ACP} \equiv \widehat{CPE}$ and $Q$ be a point on the prolongation of side $CD$, with $D \in [CQ]$,such that $\widehat{QBD} \equiv \widehat{AQB}$. Show that the triangles $PCD$ and $PBC$ are equivalent (have the same area).*

We may also consider the reciprocals of Problems 3-3e, even if they have no unique solution. The reciprocal of Problem 3 is given below:

**Problem 3g.** *Construct (using ruler and compass) a convex quadrilateral having the same area as that of a given triangle.*

**Remark.** We may search for a particular quadrilateral, as in the following problem.

**Problem 3h.** *Let $ABC$ be a given triangle. Construct using ruler and compass a trapezoid having the same area as the triangle.*

**Solution.** On one side of the triangle (see Figure 4), say $AC$, we consider an arbitrary point $C'$. We draw through $C$ a parallel $d$ to $BC'$. For any point $M \in d$ we have

$$Area[MBC'] = Area[CBC'].$$

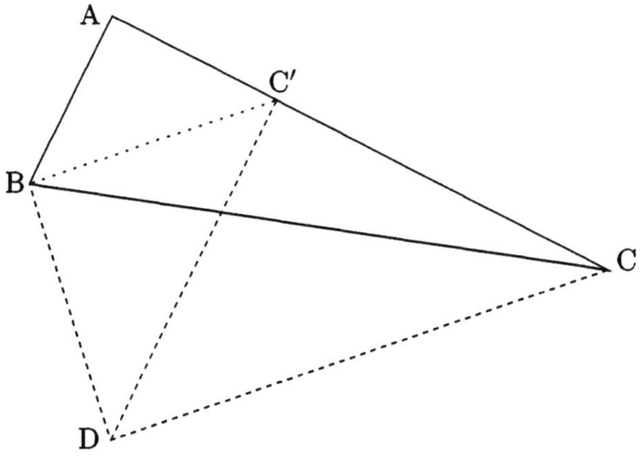


Figure 4

Therefore, in order to obtain a trapezoid, we draw through $C'$ a parallel $d'$ to $AB$ and let $\{D\} = d \cap d'$. Then $ABDC'$ is the required trapezoid. □

At the end of this paragraph we invite the reader to explore whether it is possible to construct or not (using ruler and compass) a triangle having the same area as a given concave quadrilateral.

## 3.4 Consequences. Yet More Directions

Any discovery, no matter how small it is, influences the field to which it belongs in terms of impact, reappraisals, enlightening and new directions of exploration that are thus opened.

This is why a new discovery does not close a research field; at most it explains a problem or a question and raises, by means of its influence, new questions and problems.

This is the research paradox: as much we work on a research topic, as many new research topics will be opened. We shall illustrate this "chain reaction" that is unleashed by a discovery using the most simple available example related to the topic of this section.

**Problem 3i.** *Construct a line that divides a convex quadrilateral surface into two parts having the same area.*

**Solution.** Let $ABCD$ be the given surface. We first construct a triangle having the same area as $ABCD$. Let it be $PCD$. If $E$ is the middle point of side $PD$ and $E \in (AD)$, then we have

$$Area[PCE] = Area[CED] \text{ and hence } Area[CED] = \frac{1}{2}Area[ABCD].$$

The construction is the following. We draw the diagonal $AC$, then we draw through $B$ a parallel to $AC$ and let $P$ be its intersection point to $AD$. We construct $E$, the middle point of $PD$ and then $CE$ will be just the line that divides the quadrilateral surface into two equal parts (when $E \notin (AD)$, we shall consider the other diagonal or another parallel). □

**Remark.** As we can see, our benefit was twofold: first, by using the solution of Problem 3 we solved Problem 3i; then we got another source of new problems. In the following we shall formulate only a few of them and we shall leave to the readers the satisfaction of proposing (and solving!) all the remaining ones.

**Problem 3j.** *Construct a line that divides a pentagonal surface into two equal parts.*

**Problem 3k.** *Construct a plane that divides a prism into two parts having the same volume.*

**Problem 3l.** *Construct a convex quadrilateral whose area is two times more than the area of a given convex pentagon.*

**Hint.** We first construct a triangle having the same area as the given pentagon, than we construct another triangle having the area twice that of the first one, and, finally, we construct the desired quadrilateral. □

Other branching directions of the problems in this section are suggested by problem 795 in [7].

**Problem 3m.** *Construct a square equivalent to a (convex) pentagon $ABCDE$.*

**Hint.** We first construct a triangle equivalent to the given pentagon and then use the next problem.

**Problem 3n.** *Construct a square having the same area as a given triangle.*

A very nice **solution** to Problem 3n can be found in [7], p. 279.

## 3.5 Exploring the Suggestions

It is very important for us to have found another direction for creating new problems. We state here the simplest one.

**Problem 3o.** *Construct a square having the same area as a given convex quadrilateral.*

**Hint.** We construct a triangle equivalent to a given quadrilateral (Problem 3) and then use Problem 3n.

**Problem 3p.** *Construct a regular dodecagon having the same area as a given convex quadrilateral.*

**Hint.** We use problem 796 from [7] in order to construct a regular dodecagon equivalent to a triangle (which is in turn equivalent to a given quadrilateral).

Finally, we recommend to the readers to find by themselves other related problems of this kind.

## 3.6 Bibliographical Comments

We preferred Problem 3 as the source problem in this section instead of problem 9 [5], p. 19, (which is in fact problem 794 [7] , p. 86,), whose statement requires us to construct a triangle having the same area as a given pentagon, actually the problem that has inspired the material in this section.

The topic of this section was intensively used by the author in [1] and [2] in order to illustrate the way in which a problem can be creatively solved. The paper [3] is a corollary of these papers.

An excellent detailed survey on geometric constructions may be found in [6], chapter I, section 4, pp. 52–65.

## 3.7 References

1. **V. Berinde,** *How to Stimulate the Engineering Invention by Means of Mathematical Approaches,* Proceed. of the International Conference "Teaching Mathematics for Industry" (Czech Technical University, 18-20 Sept. 1994, J. Cerny (editor), pp. 19-30

2. **V. Berinde,** *A Collection of Creatively Solved Problems which Can Develop the Inventive Skills of Students,* International Conf. on Technology in Mathematics Teaching, Napier University, Edinburgh, 4-7 Sept. 1995

3. **V. Berinde,** *A Natural Way to Initiate Students in Mathematical Research: Creative Problems Solving Sessions.* Heuristics and Didactics of Exact Sciences (Donetsk), no. 7 (1997), 30–32

4. **V. Berinde, O. Cosma,** *An elementary geometric construction performed by a computer.* Bul. Stiint. Univ. Baia Mare, **10** (1994), 99–107

5. **A. Cota et al.,** *Mathematics, Geometry and Trigonometry. Textbook for X Form.* (Romanian), Editura Didactica şi Pedagogica, 1983

6. **N. Ghircoiasu et al.,** *Fiches of Geometry and Trigonometry for Students and High School Graduates.* (Romanian), Editura Dacia, Cluj-Napoca, 1978

7. **Gh. Titeica,** *Problems of Geometry.* (Romanian), *6th* Edition, Ed. Tehnica, 1982

8. **I. Sharygin**, *Two articles and two hundred problems*, ICME9, Tokyo, 2000

# Theme 4

# Solving a Class of Nonlinear Systems

In the present section we shall be concerned with solving a typical system problem. Based on the method obtained in this case, we shall do some explorations into this class of problems in order to illustrate the generality of our method as well as to indicate the way we can discover new related problems. It is generally more difficult to tackle problems of this type using other methods than those described below.

## 4.1 Source Problem

**Problem 4.** *Solve the system:*

$$\begin{cases} \left(\frac{1}{4}\right)^x + \left(\frac{1}{27}\right)^y = \frac{5}{6}, \\ \log_{\frac{1}{4}} x - \log_{\frac{1}{27}} y = \frac{1}{6}. \end{cases}$$

## 4.2 Solving the Source Problem

**Solution 1.** Let us notice that $\frac{1}{2} + \frac{1}{3} = \frac{5}{6}$, which shows that $x = \frac{1}{2}$ and $y = \frac{1}{3}$ satisfy the first equation of the system. It is then easy to check that they also satisfy the second equation, for $\frac{1}{2} - \frac{1}{3} = \frac{1}{6}$. Therefore the pair $(x, y) = \left(\frac{1}{2}, \frac{1}{3}\right)$ is a **solution** of the system.

We shall prove it is the *unique* solution. We suppose there exists another solution with $x > \frac{1}{2}$. Then $\left(\frac{1}{4}\right)^x < \left(\frac{1}{4}\right)^{1/2} = \frac{1}{2}$ and using the first equation we have that

$$\frac{5}{6} = \left(\frac{1}{4}\right)^x + \left(\frac{1}{27}\right)^y < \frac{1}{2} + \left(\frac{1}{27}\right)^y,$$

that is

$$\left(\frac{1}{27}\right)^y > \frac{1}{3} \Leftrightarrow \left(\frac{1}{27}\right)^y > \left(\frac{1}{27}\right)^{\frac{1}{3}},$$

which, based on the monotonicity of the exponential function, leads to the conclusion that

$$y < \frac{1}{3}.$$

Therefore, if $x > 1/2$, then we necessarily have $y < 1/3$. But for $x > 1/2$ we have $\log_{\frac{1}{4}} x < \log_{\frac{1}{4}} \frac{1}{2}$ and $\log_{\frac{1}{27}} y > \log_{\frac{1}{27}} \frac{1}{3}$, since in this case the logarithmic function is decreasing. Hence

$$\log_{\frac{1}{4}} x < \log_{\frac{1}{4}} \frac{1}{2} \quad \text{and} \quad -\log_{\frac{1}{27}} y < -\log_{\frac{1}{27}} \frac{1}{3},$$

and so, by summing the previous inequalities side by side, it results that

$$\log_{\frac{1}{4}} x - \log_{\frac{1}{27}} y < \frac{1}{6},$$

a contradiction with respect to the second equation of the system. Hence there are no solutions with $x > \dfrac{1}{2}$.

In a similar manner we solve the case $x < \dfrac{1}{2}$. □

**Remark 1.** The statement of Problem 4 does not specify the set where we should search for the solutions but the context suggests that we must have $x, y \in (0, \infty)$, so that the logarithmic function is well-defined.

**Solution 2.** The system may be equivalently written as

$$\begin{cases} y = \log_{\frac{1}{27}} \left(\frac{5}{6} - \left(\frac{1}{4}\right)^x\right), \\ y = \left(\frac{1}{27}\right)^{\log_{\frac{1}{4}} x - \frac{1}{6}}, \end{cases}$$

hence we have to solve, with respect to $x$, the equation

$$\log_{\frac{1}{27}} \left(\frac{5}{6} - \left(\frac{1}{4}\right)^x\right) = \left(\frac{1}{27}\right)^{\log_{\frac{1}{4}} x - \frac{1}{6}}, \tag{1}$$

and so that is why we consider the functions in the left-hand side and right-hand side of (1), respectively.

The first function, that is

$$f(x) = \log_{\frac{1}{27}} \left(\frac{5}{6} - \left(\frac{1}{4}\right)^x\right),$$

is defined for $x \in (\log_{\frac{1}{4}} \frac{5}{6}, \infty)$ and the second one

$$g(x) = \left(\frac{1}{27}\right)^{\log_{\frac{1}{4}} x - \frac{1}{6}} \quad \text{for } x \in (0, \infty).$$

As $\alpha = \log_{\frac{1}{4}} \dfrac{5}{6} > 0$, we will look for the solution in the interval $(\alpha, \infty)$. It is easy to check that

$$g(x) = (3^{-3})^{-\log_4 x - \frac{1}{6}} = 3^{3 \log_4 x + \frac{1}{2}},$$

which shows that $g$ is strictly increasing (the bases of both logarithm and exponential in the last expression are numbers greater than 1). With regard to the function $f$, having in view the fact that $\left(\dfrac{1}{4}\right)^x$ is strictly decreasing, it results that $\dfrac{5}{6} - \left(\dfrac{1}{4}\right)^x$ is strictly increasing. As $\dfrac{1}{27} < 1$, we deduce that the function

$$f(x) = \log_{\frac{1}{27}} \left(\frac{5}{6} - \left(\frac{1}{4}\right)^x\right)$$

is strictly decreasing.

Since $f(1/2) = g(1/2)$, it results that $x = \dfrac{1}{2}$ is the unique solution of equation (1). From the first equation of the system it results that $y = \dfrac{1}{3}$ and so $\left(\dfrac{1}{2}, \dfrac{1}{3}\right)$ is the unique solution of the system. □

**Remark 2.** As we can see from the two methods of solving Problem 4, the monotonicity (of logarithmic and exponential functions) played a crucial role. Both methods are general enough to be used for solving any similar problem. By considering various pairs of elementary functions, we can obtain several new problems in this class. In the following we give a short list, most of them being proposed at mathematics competitions by different authors.

**Problem 4a.** *Solve the simultaneous equations*

$$\begin{cases} \log_3 x + \log_2 y = 2, \\ 3^x - 2^y = 23. \end{cases}$$

**Problem 4b.** *Solve the simultaneous equations*

$$\begin{cases} 3^{2x+1} + 2^{y+1} = 31, \\ \log_3(x+2) - \log_2(x+1) = 0. \end{cases}$$

**Problem 4c.** *Solve the simultaneous equations*

$$\begin{cases} 5^x + 5^y = 30, \\ 3^x - y^3 = 8. \end{cases}$$

**Problem 4d.** *Solve the simultaneous equations*

$$\begin{cases} \sqrt[3]{3x-1} - \sqrt[4]{3x-1}, \\ (x-2)^3 + (2y-1)^4 = 2. \end{cases}$$

**Problem 4e.** *Solve the system*

$$\begin{cases} 2^x + y + \log_2 z = 6 + \sqrt{2-x}, \\ 2^y + z + \log_2 x = 5 + \sqrt{1-y}, \\ 2^z + x + \log_2 y = 6 + \sqrt{2-z}. \end{cases}$$

We add to the list of problems above two other variations on the same theme and invite readers to find other interesting combinations of nonlinear monotone functions.

**Problem 4f.** *Solve the simultaneous equations*

$$\begin{cases} 4^x + 27^y = 5, \\ \log_4 x - \log_{27} y = -\dfrac{1}{6}. \end{cases}$$

**Problem 4g.** *Solve the simultaneous equations*

$$\begin{cases} \left(\frac{1}{4}\right)^x + 27^y = \frac{7}{2}, \\ \log_4 x - \log_{27} y = -\frac{5}{6}. \end{cases}$$

**Important Remark.** According to what has already become a habit of ours in tackling a problem, let us not just accept the idea that everything has been said about Problem 4 and its relatives.

If we survey the second method of solving Problem 4, no other perspective is immediately open. While if we skim the arguments involved in the first method we notice that, in the case we discussed there, i.e., $x > 1/2$ (when we also have $y < 1/3$), we contradict in fact the second equation of the system. It is still possible to get a contradiction even if the second equation is in fact an **inequality.** The difficulty that occurs in this case (which is not clearly shown by our solution, because we have not analyzed actually in detail the case $x < \dfrac{1}{2}$) is that in order to get to a contradiction we would need the same second equation of the system. But we can overcome these difficulties by considering another inequality to accomplish that.

We now examine some results of our creative investigations around Problem 4 and its solution methods.

**Problem 4h.** *Solve the system*

$$\begin{cases} \left(\frac{1}{4}\right)^x + \left(\frac{1}{27}\right)^y = \frac{5}{6}, \\ \log_{\frac{1}{4}} x - \log_{\frac{1}{27}} y \geq \frac{1}{6}, \\ 27^y - 4^x \leq 1. \end{cases}$$

**Solution.** It is easy to see that $x = \frac{1}{2}$ and $y = \frac{1}{2}$ do verify the equation and the two inequalities of the system. We shall prove that $\left(\frac{1}{2}, \frac{1}{3}\right)$ is the unique solution. We assume there exists a solution with $x > \dfrac{1}{2}$. In a similar way to Solution 1, we get that $y < \dfrac{1}{3}$. In this case

$$x > \frac{1}{2} \Longrightarrow \log_{\frac{1}{4}} x < \log_{\frac{1}{4}} \frac{1}{2} \Longleftrightarrow \log_{\frac{1}{4}} x < \frac{1}{2} \tag{2}$$

and

$$y < \frac{1}{3} \Longrightarrow \log_{\frac{1}{27}} y > \log_{\frac{1}{27}} \frac{1}{3} \Longleftrightarrow \log_{\frac{1}{27}} > \frac{1}{3} \Longleftrightarrow -\log_{\frac{1}{27}} < -\frac{1}{3}. \tag{3}$$

By summing (2) and (3) side by side, it results that

$$\log_{\frac{1}{4}} x - \log_{\frac{1}{27}} y < \frac{1}{6},$$

which contradicts the first inequality of the system. Therefore, we cannot have solutions with $x > \frac{1}{2}$. We assume there exists a solution with $x < \frac{1}{2}$. By using the equation of the system, we get $y > \frac{1}{3}$. In this case, we have

$$27^y > 27^{\frac{1}{3}} \Longleftrightarrow 27^y > 3 \tag{4}$$

and

$$4^x < 4^{\frac{1}{2}} \Longleftrightarrow 4^x < 2 \Longleftrightarrow -4^x > -2. \tag{5}$$

Now, by summing (4) and (5) side by side, we obtain

$$27^y - 4^x > 1,$$

which contradicts the second inequality of the system. □

**Remarks. 1)** By exploiting the idea of splitting an equation into two different inequalities with the same power as the initial equation, we can construct many other nonlinear systems;

**2)** It is quite clear that we can replace the remaining equation in the system by two other inequalities, whose conjoined effect should be the same as that of the equation. Now we give some brand new problems following the previous remarks and recommend to the readers to try exploring other directions, too.

One thing must be clear to everyone: we can never surely say that all latent resources of a problem have been exhausted. It is always possible to find another point of view, another starting point for further explorations.

**Problem 4i.** *Solve the system*

$$\begin{cases} 4^x + 27^y = 5, \\ \log_{27} y - \log_4 x \geq \frac{1}{6}, \\ \sqrt[3]{1-4x} + 2\sqrt[4]{6y-1} \leq 1. \end{cases}$$

**Important!** One can replace the two equations of the original system by only **three** inequalities and not four, as it is possible that their conjoined effect should be equivalent to that of the two original equations. We recommend that readers try to find by themselves such a combination. Now let us solve the next system of inequalities.

**Problem 4j.** *Solve the system*

$$\begin{cases} \left(\frac{1}{4}\right)^x + 27^y \geq \frac{7}{2}, \\ \frac{1}{x^3} + \frac{1}{y^2} \geq 17, \\ \log_{27} y - \log_4 x \leq \frac{1}{6}, \\ 4^x - \left(\frac{1}{27}\right)^y \geq \frac{5}{3}\,. \end{cases}$$

**Solution.** It is obvious that $x = \frac{1}{2}$, $y = \frac{1}{3}$ is a solution of the system. We prove its uniqueness. Let us consider a solution with $x > \dfrac{1}{2}$. It results that

$$\left(\frac{1}{4}\right)^x < \frac{1}{2} \Longleftrightarrow -\left(\frac{1}{4}\right)^x > -\frac{1}{2}\,.$$

From the first inequality we get

$$\left.\begin{array}{l} 27^y \geq \frac{7}{2} - \left(\frac{1}{4}\right)^x \\ \text{and as } -\left(\frac{1}{4}\right)^x > -\frac{1}{2} \end{array}\right\} \Longrightarrow 27^y > 3, \text{ i.e., } y > \frac{1}{3}\,.$$

In this case $\dfrac{1}{x} < 2 \Longleftrightarrow \dfrac{1}{x^3} < 8$ and $\dfrac{1}{y} < 3 \Longleftrightarrow \dfrac{1}{y^2} < 9$, which give by summation $\dfrac{1}{x^3} + \dfrac{1}{y^2} < 17$, which contradicts the second inequality of the system. Therefore we cannot have solutions with $x > \dfrac{1}{2}$.

We assume there exists a solution with $x < \dfrac{1}{2}$, then $\log_4 x < -\dfrac{1}{2}$ and from the third inequality we get

$$\log_{27}\ y \leq \log_4\ x + \frac{1}{6} < -\frac{1}{2} + \frac{1}{6}\,, \text{i.e. } \log_{27}\ y < -\frac{1}{3}$$

which leads to $y < \dfrac{1}{3}$. Hence $x < \dfrac{1}{2}$ implies $y < \dfrac{1}{3}$.

Then $4^x < 2 \Longleftrightarrow -4^x > -2$ and $\left(\dfrac{1}{27}\right)^y > \dfrac{1}{3}$, hence $\left(\dfrac{1}{27}\right)^y - 4^x > -\dfrac{5}{3}$, which contradicts the last inequality of the system. □

**Remark.** If we followed all exploratory directions suggested by our creative solvution of this kind of problems, we could write a book only on only one class of them! We are not in the position to do so and therefore limit ourselves to indicate only two other ways for obtaining new related problems:

**1)** by interchanging the signs "$\leq$ " and "$\geq$ " in a pair of inequalities;

**2)** by introducing a third unknown, as suggested by Problem 4d.

Following the previous remark 1), we can reformulate Problems 4h and 4i as follows:

**Problem 4k.** *Solve the system*

$$\begin{cases} \left(\frac{1}{4}\right)^x + \left(\frac{1}{27}\right)^y = \frac{5}{6}, \\ \log_{\frac{1}{4}} x - \log_{\frac{1}{27}} y \leq \frac{1}{6}, \\ 27^y - 4^x \geq 1. \end{cases}$$

**Problem 4l.** *Solve the system*

$$\begin{cases} 4^x + 27^y = 5, \\ \log_{27} y - \log_4 x \leq \frac{1}{6}, \\ \sqrt[3]{1-4x} + 2\sqrt[3]{6y-1} \geq 1. \end{cases}$$

## 4.3 Other Directions

Even though our investigations around Problem 4 have covered a very limited area of novelty, we however obtained some interesting new problems as for instance Problems 4h – 4k. In order to obtain them, we exploited mainly the arguments in Solution 1 of Problem 4. This solution has allowed us to detect the sufficient role of the sign "=" in the architecture of our systems.

We want to show now that, by adapting Solution 2 of Problem 4, we can also solve systems that contain inequalities.

Let us look at Problem 4h. We need to prove that $\left(\dfrac{1}{2}, \dfrac{1}{3}\right)$ is the unique solution of this system. From the equation we get

$$y = \log_{\frac{1}{27}} \left[\frac{5}{6} - \left(\frac{1}{4}\right)^x\right],$$

while from the two inequalities we get

$$y \geq \left(\frac{1}{27}\right)^{\log_{\frac{1}{4}} x - \frac{1}{6}},$$

and

$$y \leq \log_{27}(1 + 4^x),$$

respectively.

Therefore $x$ is the solution of the double inequality

$$\left(\frac{1}{27}\right)^{\log_{\frac{1}{4}} x - \frac{1}{6}} \leq \log_{\frac{1}{27}} \left[\frac{5}{6} - \left(\frac{1}{4}\right)^x\right] \leq \log_{27}(1 + 4^x). \tag{6}$$

Let us denote by $f, g$ and $h$ the expressions involved in (6), in this order, which are defined on $(\alpha, \infty)$,

$$\alpha = \log_{\frac{1}{4}} \frac{5}{6} > 0, \quad \alpha < \frac{1}{2}.$$

It is an easy task to show that $f$ and $h$ are strictly increasing, while $g$ is strictly decreasing.

Since $f\left(\frac{1}{2}\right) = g\left(\frac{1}{2}\right) = h\left(\frac{1}{2}\right)$, the inequalities (6), i.e., $f(x) \leq g(x) \leq h(x)$, simultaneously hold if and only if $x = \frac{1}{2}$ (see Figure 5).

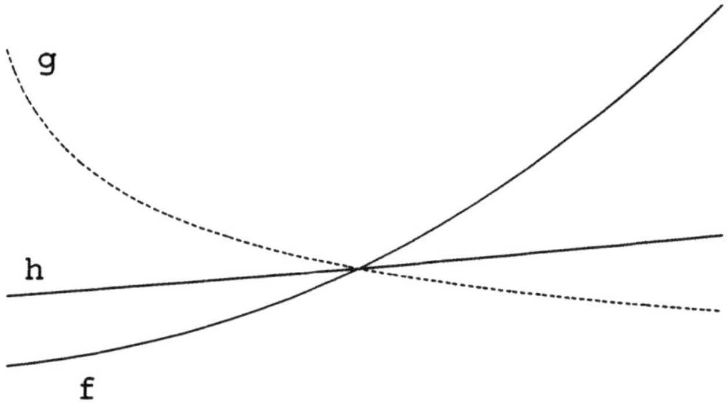


Figure 5

**Remark 3.** In fact the graph argument reduces to the algebraic one we used: if $x > \frac{1}{2}$, then $f(x) > f\left(\frac{1}{2}\right) = g\left(\frac{1}{2}\right) > g(x)$, which is a contradiction, and so on.

**A New Discovery.** It is now clear that we have just found another source of new problems, i.e., inequalities of the form (6). Let us formulate two problems of this kind.

**Problem 4l$'$.** *Solve the simultaneous inequalities*

$$\log_{27}(1 + 4^x) \leq \log_{\frac{1}{27}} \left[\frac{5}{6} - \left(\frac{5}{6}\right)^x\right] \leq \log_{\frac{1}{4}} x - \frac{1}{6}.$$

**Problem 4m.** *Find the range of values of $x$ for which*

$$\log_{\frac{1}{27}} \left[\frac{5}{6} - \left(\frac{5}{6}\right)^x\right] \leq \log_{27}(1 + 4^x).$$

**Hint.** The above arguments suggest that $x \in [1/2, \infty)$ is the required solution. □

The main benefit of our last creative approach is the statement of the following general principle for solving inequality systems of the form (6).

Let $f, g, h : I \to \mathbb{R}$ ($I =$ interval) be three functions satisfying

1) there exists $x_0 \in I$ such that $f(x_0) \leq g(x_0) \leq h(x_0)$;

2) $f, g$ and $h$ are strictly monotone, $g$ being of opposite monotonicity to that of $f$ and $h$.

Then the system of inequalities

$$f(x) \le g(x) \le h(x)$$

does have the unique solution $x = x_0$.

**Problem 4n.** *Solve each of the following systems*

$$a) \begin{cases} \log_3 x + \log_5 y = 2, \\ 3^y - 5^x = 118, \end{cases} \qquad b) \begin{cases} \log_2(1+\sqrt{x}) - \log_3 y = 0, \\ y^2 + 2^x = 1, \end{cases}$$

$$c) \begin{cases} \log_3 x + \log_5 y = 2, \\ 3^y - 5^x \ge 118, \\ \log_{\frac{1}{3}} x - \log_{\frac{1}{5}} y \le -2, \end{cases} \qquad d) \begin{cases} y^2 + 2^x = 11, \\ \log_2(1+\sqrt{x}) \ge \log_3 y, \\ 5^{\log_2 2x} + \left(\frac{1}{3}\right)^{y-2} \le \frac{16}{3}, \end{cases}$$

$$e) \ \left(\tfrac{1}{5}\right)^{\log_{\frac{1}{3}} x+2} \le 5^{2-\log_3 x} \le \log_3(5^x + 118),$$

$$f) \ 2 + \log_{\frac{1}{3}} \left[\tfrac{16}{3} - 5^{\log_2 2x}\right] \le \sqrt{11 - 2^x} \le 3^{\log_2(1+\sqrt{x})}.$$

**Important Remark.** If $x_0$ is an extremity of the interval $I$, then it is possible to get a unique solution even if we have only one inequality instead of a system containing two or more inequalities, as in the following problem.

**Problem 4o.** *Find the range of values of $x$ for which*

$$2^{3+\sqrt{x-2}} + 3^x \le \log_3(85 - 2^x) + \left(\frac{1}{3}\right)^{x-4} + 4.$$

**Solution.** Here we have

$$f : [2, \infty) \to \mathbb{R}, \quad f(x) = 2^{3+\sqrt{x-2}} + 3^x$$

which is an increasing function, while

$$g : (0, \log_2 85) \to \mathbb{R}, \ g(x) = \log_3(85 - 2^x) + \left(\frac{1}{3}\right)^{x-4} + 4$$

is decreasing. The inequality is defined on the interval $[2, \log_2 85)$.

As $f(2) = g(2)$, $f(x) \le g(x)$, $f$ is increasing and $g$ is decreasing, it results that $x = 2$ is the unique solution. □

**Problem 4p.** *Find the range of values of $x$ for which*

$$3^x + 4^x + 9 \cdot 5^x \ge 15^x + 5^x 2^{\sqrt{x-2}}.$$

**Solution.** The inequality may be written in the equivalent form

$$3^x + 4^x \ge 5^x \left[2^{\sqrt{x-2}} + 3^x - 9\right]$$

which becomes, dividing it side by side by $5^x > 0$,

$$\left(\frac{3}{5}\right)^x + \left(\frac{4}{5}\right)^x \geq 2^{\sqrt{x-2}} + 3^x - 9\,.$$

As $x = 2$ is a solution, $f(x) \geq g(x)$, $\forall\, x \in [2, \infty)$, $f$ is strictly decreasing and $g$ strictly increasing, we have that $x = 2$ is the unique solution of the inequality. □

Other problems related to the ones in this section are those of the following type, which are presented in a more detailed manner in Theme 10 of this book.

**Problem 4r.** *Solve in* $\mathbb{R}$ *the system*

$$\begin{bmatrix} 3^x + 4^x = 5^y, \\ 3^y + 4^y = 5^z, \\ 3^z + 4^z = 5^x. \end{bmatrix}$$

**Problem 4s.** *Solve in* $\mathbb{R}$ *the system*

$$\begin{bmatrix} 3^x + 4^y = 5^z, \\ 3^y + 4^z = 5^x, \\ 3^z + 4^x = 5^y. \end{bmatrix}$$

**Problem 4t.** *Solve in* $\mathbb{R}$ *the system*

$$\begin{bmatrix} 3^x + 4^y = 5^x, \\ 3^y + 4^z = 5^y, \\ 3^z + 4^x = 5^z. \end{bmatrix}$$

**Solution of Problem 4s.** (We present here the solution A in [2].)

We sum up the equations of the system 4s (in the same way as for Problem 4r) and we obtain

$$(3^x + 4^x - 5^x) + (3^y + 4^y - 5^y) + (3^z + 4^z - 5^z) = 0. \tag{7}$$

Considering the property of monotonicity, we can state that the equation $3^x + 4^x = 5^x$ has $x = 2$ as a unique solution. Thus from (7) we have that $x = y = z = 2$ is a solution of the system. The next step is to show this is the unique solution. We assume that there is another solution with $x > 2$.

From the first equation we obtain

$$4^y = 3^x[(5/3)^x - 1]$$

and because the function $f : [0, \infty) \to \mathbb{R}$, $f(x) = 3^x[(5/3)^x - 1]$ is strictly increasing (as a product of strictly increasing functions) we deduce that

$$x > 2 \Longrightarrow f(x) > 16, \text{ i.e., } 4^y > 16, \text{ so } y > 2.$$

Similarly, we get $z > 2$, which contradicts (7), because for these values $3^x + 4^x - 5^x < 0$, $3^y + 4^y - 5^y < 0$, $3^z + 4^z - 5^z < 0$. We follow the same method for $x \in (0, 2)$ and separately for $x \leq 0$. □

## 4.4 Bibliographical Comments

Problem 4, which started the cascade of explorations in this section, was proposed by V. Chiriac in *Gazeta Matematica* no. 6/1986 as Problem 20797*. Problems 4a – 4e are variations on the same theme by different authors and have been proposed at several mathematics contests. Thus, Problem 4a was proposed by I.V. Maftei and Gh. Pantelimon for the county round of the National Mathematics Olympiad in 1986. By that time, the questions were unique nationwide.

Problem 4b was proposed by L. Toth for the Intercounty Mathematics Contest "Gr.C. Moisil" in 1989 and Problem 4e was proposed by N. Pavelescu for the Intercounty Mathematics Contest "N. Paun" in 1996. We don't know the authors of the other problems: Problem 4c was given at the Intercounty Mathematics Contest "Spiru Haret - Gh. Vranceanu" in 1991, while Problem 4e was proposed at the Intercounty Mathematics Contest "Tr. Lalescu" in 1992.

Problem 4h, proposed by the author of the book, was a question in the National Mathematics Olympiad, final round, Alba Iulia, 1999.

Problem 4s was proposed in 1993 by the same author for the Intercounty Mathematics Contest "Gr.C. Moisil" in Zalau (see [1]), while Problem 4r is actually problem C:1878 from *Gazeta Matematica* no. 12/1996.

Digging around these problems — and similar ones — shows how important mathematics contests are in keeping a continuous emulation and a remarkable development process in creating mathematical problems in our country. Great efforts must be made in order to preserve these traditions, in spite of the financial restrictions which have affected the Romanian education system in the last 10–11 years.

## 4.5 References


1. **D. Andrica, V. Berinde et al.,** *The Intercounty Mathematics Competition "Gr.C. Moisil". Editions I-XV (1986-2000).* (Romanian), Editura Cub Press 22, 2001

2. **V. Berinde,** *Two Solutions of a Competition Problem,* Mathematics Competitions, **11** (1998), no. 1, 36–44

3. **O. Cojocaru,** *The Intercounty Mathematics Competition "Spiru Haret-Gh. Vrânceanu". 1985-1966. High School Level.* (Romanian), Editura Paralela 45, 1998

4. **M. Neacsu, C. Neacsu, C.M. Semenescu,** *The Intercounty Mathematics Competition "Traian Lalescu". Editions I-XI.* (Romanian), Editura Timpul, 1997

5. —, The Collection of *Gazeta Matematica* (1980–2000)

# Theme 5

# A Class of Homogenous Inequalities

In this section we will confine our study to some inequalities which have a special form and were deduced by combining two problems, both considered as source problems of the theme.

## 5.1 Source Problems

**Problem 5.** Let $a, b > 0$. *Prove that for any* $n \in \mathbb{N}^*$,

$$\left(\frac{a+b}{2}\right)^n \leq \frac{a^n + a^{n-1}b + \ldots + ab^{n-1} + b^n}{n+1}.$$

**Problem 5′.** *If* $a, b \geq 0$, *then*

$$\left(\frac{a+b}{2}\right)^n \leq \frac{a^n + b^n}{2}.$$

## 5.2 Solving the Source Problems

We are going to use the mathematical induction method.

**Solution of Problem 5.** For $n = 1$ we have equality, and for $n = 2$ the inequality $\frac{a^2 + 2ab + b^2}{4} \geq \frac{a^2 + ab + b^2}{3}$ is equivalent to $(a - b)^2 \geq 0$, which is obvious. In both of these cases, the equality occurs if and only if $a = b$.

We assume now that the inequality holds for $k$, that is

$$\left(\frac{a+b}{2}\right)^k \leq \frac{a^k + a^{k-1}b + \ldots + ab^{k-1} + b^k}{k+1} \tag{1}$$

and prove that it also holds for $k+1$, that is

$$\left(\frac{a+b}{2}\right)^{k+1} \leq \frac{a^{k+1}+a^k b+\ldots+ab^k+b^{k+1}}{k+2}. \tag{2}$$

We therefore multiply both sides of (1) by $\frac{a+b}{2} > 0$, thus obtaining

$$\left(\frac{a+b}{2}\right)^{k+1} \leq \frac{(a+b)(a^k+a^{k-1}b+\ldots+b^k)}{2(k+1)}.$$

In order to prove (2) it is enough to show that

$$\frac{(a+b)(a^k+a^{k-1}b+\ldots+b^k)}{2(k+1)} \leq \frac{a^{k+1}+a^k b+\ldots+ab^k+b^{k+1}}{k+2}.$$

This last one is successively equivalent to

$$\begin{aligned}
&(k+2)[a^{k+1}+2(a^k b+a^{k-1}b^2+\ldots+ab^k)+b^{k+1}] \\
&\leq (2k+2)(a^{k+1}+a^k b+\ldots+b^{k+1}) \\
&\Leftrightarrow 2(a^k b+a^{k-1}b^2+\ldots+ab^k) \leq k(a^{k+1}+b^{k+1}) \\
&\Leftrightarrow (ka^{k+1}-a^k b-a^{k-1}b^2-\ldots-ab^k) \\
&\quad +(kb^{k+1}-a^k b-a^{k-1}b^2-\ldots-ab^k) \geq 0.
\end{aligned} \tag{3}$$

By grouping the terms in each of the parenthesis above we obtain

$$ka^{k+1}-a^k b-a^{k-1}b^2-\ldots-ab^k = a^k(a-b)+a^{k-1}(a^2-b^2)+\ldots+a(a^k-b^k)$$

and, respectively,

$$kb^{k+1}-a^k b-a^{k-1}b^2-\ldots-ab^k = b^k(a-b)+b^{k-1}(a^2-b^2)+\ldots+b(a^k-b^k)$$

and so (3) becomes:

$$(a-b)(a^k-b^k)+(a^2-b^2)(a^{k-1}-b^{k-1})+\ldots+(a^k-b^k)(a-b) \geq 0$$

which obviously holds, since

$$(a^i-b^i)(a^j-b^j) = (a-b)^2 \cdot E \geq 0, \tag{4}$$

for any $i, j \in \mathbb{N}^*$ and $a, b > 0$ (E involved in (4) is an expression greater than zero).
The equality holds if and only if $a = b$. □

## 5.3 A Creative Approach to Problem 5 and Problem 5′

The first thing to be noticed is that the left side is the same in both source inequalities. The idea that naturally appears now is to compare the expressions in the right-hand side of these inequalities, i.e., to find out which of the numbers

$$\frac{a^n+b^n}{2} \text{ and } \frac{a^n+a^{n-1}b+\ldots+ab^{n-1}+b^n}{n+1}$$

is greater. Thus we obtain

**Problem 5a.** *If* $a, b \geq 0$, *then*

$$\frac{a^n + a^{n-1}b + \ldots + ab^{n-1} + b^n}{n+1} \leq \frac{a^n + b^n}{2}.$$

**Solution.** We prove this directly by grouping. We get

$$\begin{aligned}
&2(a^n + a^{n-1}b + \ldots + ab^{n-1} + b^n) \leq (n+1)(a^n + b^n) \\
&\Leftrightarrow (n-1)a^n - 2a^{n-1}b - \ldots - 2ab^{n-1} + (n-1)b^n \geq 0 \\
&\Leftrightarrow (a-b)(a^{n-1} - b^{n-1}) + (a^2 - b^2)(a^{n-2} - b^{n-2}) + \ldots \\
&\quad +(a^{n-1} - b^{n-1})(a-b) \geq 0,
\end{aligned}$$

which is true according to (4). The equality holds if and only if $a = b$. □

**Remarks.** 1) Problem 5a shows that the inequality in Problem 5 is stronger than the one in Problem 5′, more precisely, Problem 5′ is a consequence of Problems 5 and 5a;

2) As the title suggests, the inequalities studied in this section are homogenous, i.e., all the terms in a certain inequality have the same order. This suggests us the following extension:

**Problem 5b.** *Let* $a, b > 0$, $n \in \mathbb{N}^*$ *and the numbers* $k_0, k_1, \ldots, k_n$; $l_0, l_1, \ldots, l_n$ *so that* $k_i + l_{n-i} = n$, *for any* $i = 0, 1, 2, \ldots, n$. *Then*

$$\frac{a^n + b^n}{2} \geq \frac{a^{k_0}b^{l_n} + a^{k_1}b^{l_{n-1}} + \ldots + a^{k_n}b^{l_0}}{n+1}$$

*and the equality holds if and only if* $a = b$.

The solution of this problem is similar to that of the particular cases solved above and therefore will not be given here explicitly. The first consequence of Problem 5b is the following one, which stands as a generalization of the first source problem.

**Problem 5c.** *Provided that all the conditions in Problem 5b are satisfied, we have*

$$\frac{a^{k_0}b^{l_n} + a^{k_1}b^{l_{n-1}} + \ldots + a^{k_n}b^{l_0}}{n+1} \geq \left(\frac{a+b}{2}\right)^n.$$

**Remark.** Obviously, for $k_i = n - i = l_i$; $i = 0, 1, 2, \ldots, n$, from Problem 5b we get Problem 5a, while from Problem 5c we obtain Problem 5.

Below we give some remarkable particular cases of Problems 5b and 5c.

**Problem 5d.** *If* $a, b > 0$, *then*

$$\frac{a^3 + b^3}{2} \geq \frac{a^2b + ab\sqrt{ab} + ab^2}{3}.$$

**Solution.** The inequality above can be written as

$$\frac{a^3+b^3}{2} \geq \frac{a^2b^1 + a\sqrt{a}\cdot b\sqrt{b} + a^1b^2}{3},$$

which is equivalent to

$$2(a-b)(a^2-b^2) + \left(a\sqrt{a} - b\sqrt{b}\right)^2 \geq 0,$$

obviously true. The equality holds if and only if $a = b$. □

**Problem 5e.** *If* $a, b > 0$, *then*

$$\frac{a^5+b^5}{2} \geq \frac{a^2b^3 + a^3b\,\sqrt[3]{a^2b} + a^2b^2\sqrt{ab} + ab^3\,\sqrt[3]{ab^2} + a^3b^2}{5}.$$

**Solution.** Using the same method as above and grouping the terms in a convenient way, we obtain

$$2(a^2-b^2)(a^3-b^3) + 2\left(a^{\frac{4}{3}} - b^{\frac{4}{3}}\right)\left(a^{\frac{11}{3}} - b^{\frac{11}{3}}\right) + \left(a^{\frac{5}{2}} - b^{\frac{5}{2}}\right)^2 \geq 0,$$

which is obviously true. The equality holds if and only if $a = b$. □

**Remark.** The technique we used here is not the only possible method, and we recommend also the solutions presented for the Problem 5e in *Gazeta Matematica*, no. 6/1995. The first one, given by Al. Constantinescu, is basically the same as that given above, but much simplified by using the notation $x = \sqrt[6]{a},\ y = \sqrt[6]{b},\ x = ty$. The second solution there uses some knowledge of differential calculus.

**Problem 5f.** *If* $a, b > 0$, *then*

$$\frac{a^5+b^5}{2} \geq \frac{a^4\,\sqrt[4]{ab^3} + a^3b\,\sqrt[3]{ab^2} + a^2b^2\sqrt{ab} + ab^3\,\sqrt[3]{a^2b} + b^4\,\sqrt[4]{a^3b}}{5}.$$

**Problem 5g.** *If* $a, b > 0,\ a \neq 2b,\ b \neq 2a$, *then*

$$(a^n+b^n)(2^n-2) \geq \frac{a\cdot(2b)^n - 2ba^n}{2b-a} + \frac{b\cdot(2a)^n - 2ab^n}{2a-b}, \quad \forall\, n \in \mathbb{N}^*.$$

*When does the equality hold ?*

**Solution.** We have

$$\begin{aligned}\frac{a\cdot(2b)^n - 2ba^n}{2b-a} &= \frac{a[(2b)^n - a^n] - (2b-a)a^n}{2b-a}\\ &= a[(2b)^{n-1} + (2b)^{n-2} + \ldots + 2b\cdot a^{n-2} + a^{n-1}] - a^n.\end{aligned}$$

Similarly

$$\frac{b\cdot(2a)^n - 2a\cdot b^n}{2a-b} = b[(2a)^{n-1} + \ldots + 2a\cdot b^{n-2} + b^{n-1}] - b^n$$

and the inequality to be proved becomes

$$(a^n + b^n)(2^n - 1) \geq a[(2b)^{n-1} + \ldots + a^{n-1}] + b[(2a)^{n-1} + \ldots + b^{n-1}].$$

One can notice that the sum of the coefficients of the terms in each of the brackets on the right-hand side is $2^{n-1} + 2^{n-2} + \ldots + 2 + 1 = 2^n - 1$. We are thus lead to the idea of moving all the expressions on the left side and, by grouping, we will obtain a sum of products of the form

$$A_{ik} = (a^i - b^i)(a^k - b^k), \quad i, k \in \mathbb{N}^*.$$

This sum is obviously $\geq 0$ ( since each $A_{ik} \geq 0$).

The equality holds if and only if all $A_{ik}$ are zero, i.e., $a = b$. □

**Problem 5h.** *If* $a + b > 0$, *then*

$$\frac{a^5 + b^5}{a^4 + a^3 b + a^2 b^2 + ab^3 + b^4} \geq \frac{a + b}{5}.$$

**Solution.** We will first prove that $a^4 + a^3 b + a^2 b^2 + ab^3 + b^4 > 0$, provided that $a + b > 0$. If $a = 0$, then $b > 0$ and obviously $b^4 > 0$. Similarly, in the case $b = 0$ and $a > 0$. We assume now that $a \cdot b \neq 0$ and, dividing by $b^4 > 0$ and introducing the notation $a = tb$, the inequality above becomes

$$t^4 + t^3 + t^2 + t + 1 > 0 \Leftrightarrow (t + 1)^2(t^2 - t + 1) + t^2 > 0,$$

which holds for all $t \in \mathbb{R}$, since $t^2 - t + 1 > 0$ and $(t + 1)^2$ and $t^2$ do not vanish simultaneously. Taking into account this result and the formula

$$a^5 + b^5 = (a + b)(a^4 - a^3 b + a^2 b^2 - ab^3 + b^4),$$

the inequality in Problem 5h is equivalent to

$$4a^4 - 6a^3 b + 4a^2 b^2 - 6ab^3 + 4b^4 \geq 0$$

(we have also used the fact that $a + b > 0$). By grouping the terms in a convenient manner, the inequality above becomes

$$2(a - b)^2(2a^2 - ab + 2b^2) \geq 0,$$

which is true, since

$$2a^2 - ab + 2b^2 = (a^2 - ab + b^2) + (a^2 + b^2) > 0.$$

The equality holds if and only if $a = b$. □

**Problem 5i.** *If* $a + b > 0$, *then*

$$\frac{a^{2n+1} + b^{2n+1}}{a^{2n} + a^{2n-1} b + \ldots + ab^{2n-1} + b^{2n}} \geq \frac{a + b}{2n + 1}, \quad \forall\, n \in \mathbb{N}^*.$$

**Remark.** For $n = 2$, Problem 5i reduces to Problem 5h.

## 5.4 Bibliographical Comments

Problem 5 of this section is exactly Problem C:454 in *Gazeta Matematica* no. 11/1984, having M. Lascu and C. Neacsu as authors, while Problem 5′ is Problem 31 in [1]. For $n = 3$, Problem 5 leads us to Problem E:8403 in *Gazeta Matematica* no. 10/1984, proposed by Radu Maier.

Problem 5d appears in *Gazeta Matematica* no. 8/1994 as E:10769, proposed by the author of this book, and it was also one of the questions at the Intercounty Mathematics Conpetition "Gr. C. Moisil" in 1995 and at the county round of the National Mathematics Olympiad 1996 in Maramures.

Problem 5e was published in *Gazeta Matematica*, no. 9/1994 as 23059 and is solved in *Gazeta Matematica*, no. 6/1995, pages 271–272.

Problem 5g was one of the questions at the Intercounty Mathematics Competition "Gr.C. Moisil", the 14th edition, Baia Mare, 26–28 March 1999 and was also published in *Gazeta Matematica*, no. 11/1989.

Problem 5h was published in *Gazeta Matematica* no. 9/1987 as problem OG:55, having L. Tutescu as author.

## 5.5 References

1. **E. Rogai,** *Tables and Mathematical Formulas.* (Romanian), Editura Tehnica, 1984

2. **Gh. Andrei,** *Exercises and Problems in Algebra for School Contests and Olympiads.* (Romanian), Constanta, 1990

3. **L. Panaitopol, V. Bandila, M. Lascu,** *Inequalities.* (Romanian), Editura GIL, 1996

4. **D.S. Mitrinovic**, *Analytic Inequalities.* Springer-Verlag, 1970.

# Theme 6

# The First Decimal of Some Irrational Numbers

In this section, after having solved the source problem, we will try to exploit the generality of the solution and to find some of its advantages. We will also try to show how important the *algorithmicity* and the *generality* are among the attributes of a creative solving method. These are both essential qualities, that can suggest many directions of exploring around the given problem.

## 6.1 Source Problem

**Problem 6.** *Find the first decimal of the number*

$$N = \sqrt{n^2 + 11n + 30}\ , \quad n \in \mathbb{N},\ n \geq 3.$$

## 6.2 Solving the Source Problem

**Solution 1.** (From *Gazeta Matematica*).
Denoting $x = n + 5$, it results that $N = \sqrt{x^2 + x}$. As

$$x^2 + x < x^2 + x + \frac{1}{4} = \left(x + \frac{1}{2}\right)^2,$$

we get $N < x + \dfrac{5}{10}$.

It is natural to check if 4 is the wanted decimal, i.e., $N > x + \dfrac{4}{10}$. The relation becomes

$$x^2 + x > \left(x + \frac{4}{10}\right)^2 \Leftrightarrow x^2 + x > x^2 + \frac{4}{5}x + \frac{4}{25} \Leftrightarrow x > \frac{4}{5}$$

and the last inequality is satisfied because $x = n+5 \geq 3+5 = 8$. As a conclusion, the first decimal of the number $N$ is 4. $\square$

**Solution 2.** Notice that

$$(n+5)^2 = n^2 + 10n + 25 < n^2 + 11n + 30 < n^2 + 12n + 36 = (n+6)^2$$

and so

$$n+5 < \sqrt{n^2+11n+30} < n+6$$

which shows that the integer part of the $N$ is $n+5$ :

$$\left[\sqrt{n^2+11n+30}\right] = n+5.$$

This suggests that $N$ is irrational. Let $a$ be its first decimal, i.e.,

$$\sqrt{n^2+11n+30} - (n+5) = 0.a\ldots$$

which could also be written in the form

$$10\left(\sqrt{n^2+11n+30} - n - 5\right) = a.b\ldots.$$

This last relation shows that $a$ is actually the integer part of the number

$$x = 10\left(\sqrt{n^2+11n+30} - n - 5\right),$$

that is

$$a \leq 10\left(\sqrt{n^2+11n+30} - n - 5\right) < a+1, \quad \forall\, n \in \mathbb{N}$$

which is equivalent to

$$\frac{a}{10} \leq \sqrt{n^2+11n+30} - n - 5 < \frac{a+1}{10}, \quad \forall\, n \in \mathbb{N},$$

i.e.,

$$n+5+\frac{a}{10} \leq \sqrt{n^2+11n+30} < n+5+\frac{a+1}{10}, \quad \forall\, n \in \mathbb{N}.$$

By squaring on both sides we obtain

$$n^2 + 2\left(5+\frac{a}{10}\right)n + \left(5+\frac{a}{10}\right)^2 \leq n^2 + 11n + 30$$

$$< n^2 + 2\left(5+\frac{a+1}{10}\right)n + \left(5+\frac{a+1}{10}\right)^2, \forall\, n \in \mathbb{N}. \tag{1}$$

A **necessary** condition for (1) to hold is that the next relations between the coefficients of $n$ in the three terms of (1) hold:

$$2\left(5+\frac{a}{10}\right) \leq 11 \leq 2\left(5+\frac{a+1}{10}\right),$$

otherwise (1) couldn't be fulfilled for any given $n \in \mathbb{N}$.

The above relations lead to the double inequality $a \leq 5 \leq a+1$, satisfied for $a \in \{4,5\}$, but only $a = 4$ can be accepted. Indeed, this value of $a$ is **sufficient** as well for (1) to hold, since $30 < \left(5+\frac{1}{2}\right)^2$. Therefore the first decimal of the irrational number $N$ is 4, for any $n \in \mathbb{N}$ (and not only for $n \geq 3$, as the statement of Problem 6 asserts). □

**Remark.** Notice that the number $N$ in Problem 6 is actually *always* irrational. Indeed, denoting $k = n+5$ we have $k(k+1) = N^2$ which is impossible for $k$ and $N$ nonzero natural numbers.

## 6.3 A Close Examination of the Solution. Exploring Around It

Notice that Solution 2 is logically constructed, leaving no impression of improvising or of artificial statements, like Solution 1. Moreover, it stresses the following fact: the double inequality

$$n+5 < \sqrt{n^2+11n+b} < n+6, \forall n \in \mathbb{N},$$

equivalent to

$$n^2+10n+25 < n^2+11n+b < n^2+12n+36, \forall n \in \mathbb{N}$$

is true for any number

$$b \in \{26, 27, ..., 35\}. \tag{2}$$

Thus, for all the values of $b$ which satisfy (2), we still have

$$\left[\sqrt{n^2+11n+b}\right] = n+5.$$

This fact immediately raises the following question: if $b$ does satisfy (2), what can we say about the first decimal of the number

$$N_b = \sqrt{n^2+11n+b} \quad ?$$

The answer, which will be obtained after following the pattern in Solution 2, isn't what we would expect. Of all the numbers $N_b$, with $b$ satisfying (2), only the one obtained for $b = 30$, which is exactly the one in Problem 6, has a constant first decimal for any $n \in \mathbb{N}$, (and this digit is 4).

This remark enables us to formulate the following new problem.

**Problem 6a.** *Find the first decimal of the number*

$$\sqrt{n^2+9n+20}\ , \quad n \in \mathbb{N}.$$

**Solution.** In a similar manner as before, we can show that, for any $b \in \{17, 18, \ldots, 24\}$, the integer part of the number $\sqrt{n^2+9n+b}$ is $n+5$. Of all these, generally irrational numbers, only the one obtained for $b = 20$, i.e., $\sqrt{n^2+9n+20}$, has the same first decimal, for any $n \in \mathbb{N}$. This digit is 4 again. □

**Remarks.** 1) From the previous considerations it results that the first decimal of the numbers $\sqrt{n^2+11n+30}$ and $\sqrt{n^2+9n+20}$ is 4, for any $n \in \mathbb{N}$, and **not only** for $n \geq 3$, as the source problem was asking.

2) Moreover, for some values of $b$ the other numbers $N_b$ have the same first digit as well, but only *beginning* from a certain rank, as shown in the next problem.

**Problem 6b.** *Find the first decimal of the number*

$$\sqrt{n^2+11n+26}, \quad n \in \mathbb{N},\ n \geq 16.$$

**Solution.** We must have

$$n+5+\frac{a}{10} \leq \sqrt{n^2+11n+26} < 5+\frac{a+1}{10},$$

which holds for $a = 4$ and any $n \geq 16$. □

**Problem 6c.** *Find $n_0$ such that the first decimal of the number*

$$\sqrt{n^2+11n+29}$$

*is the same for any $n \in \mathbb{N},\ n \geq n_0$.*

**Answer.** We obtain $n_0 = 1$ and the first decimal is $a = 4$. □

**Remarks.** 1) Continuing this pattern, for $\sqrt{n^2+11n+28}$ we get $n_0 = 3$, for $\sqrt{n^2+11n+27}$ we have $n_0 = 11$, while for $\sqrt{n^2+11n+26}$ we obtain $n_0 = 16$, as in Problem 6b.

2) Continuing the investigations, we find that the numbers

$$N_b = \sqrt{n^2+7n+b} \ \text{ with } \ b \in \{10, 11, ..., 15\}$$

have $[N_b] = n+3$ and of all these numbers only $N_{12}$ (i.e., $\sqrt{n^2+7n+12}$) has the same first decimal $a$ for any $n \in \mathbb{N}$ (and this digit is $a = 4$).

In the same manner, for the numbers $N_b = \sqrt{n^2+5n+b}$ with $b \in \{5, 6, 7, 8\}$ we have $[N_b] = n+2$, and only $N_6 = \sqrt{n^2+5n+6}$ has the first decimal the same, for any $n \in \mathbb{N}$. The investigations above may be summarized as follows.

**Problem 6d.** *Show that the integer part of the number*

$$N_b = \sqrt{n^2+(2p+1)n+b_p}$$

*is $n+p, \forall n, p \in \mathbb{N}^*$, where $b_p$ is a natural number in the set*

$$\{p^2+1,\ p^2+2, ..., p^2+2p\}.$$

*Then find $b_p$ so that the first decimal of the number $N_p$ is the same for any $n \in \mathbb{N}, n \geq 3$.*

**Solution.** The first part follows immediately, because

$$(n+p)^2 = n^2 + 2pn + p^2 < n^2 + (2p+1)n + b_p$$
$$< n^2 + (2p+2)n + p^2 + 2p + 1 = (n+p+1)^2.$$

For the second part we use the same method as in solving Problem 6. Let $a$ be the first decimal digit of the number $N_p$. Since $[N_p] = n + p$, we must have

$$\frac{a}{10} \leq \sqrt{n^2 + (2p+1)n + b_p} - n - p < \frac{a+1}{10} \tag{3}$$

for any $n \in \mathbb{N}, n \geq 3$. Thus, for $b_p = p^2 + p$ , (3) will become

$$\left(n + p + \frac{a}{10}\right)^2 \leq n^2 + (2p+1)n + p^2 + p < \left(n + p + \frac{a+1}{10}\right)^2$$

which is equivalent to

$$\frac{a}{5}n + \frac{a}{5}p + \left(\frac{a}{10}\right)^2 \leq n + p < \frac{a+1}{5} \cdot n + \frac{a+1}{5}p + \left(\frac{a+1}{10}\right)^2. \tag{4}$$

For (4) to hold for any $n$ (maybe starting from a certain rank) it is necessary that $\frac{a}{5} \leq 1 \leq \frac{a+1}{5}$, which is verified for $a \in \{4, 5\}$, but (4) holds only for $a = 4$.

In this case, the first inequality from (4) is obviously true, and so is the inequality on the right-hand side.

If $b_p > p^2 + p$, then $b_p \geq p^2 + p + 1$ and because

$$\left(p + \frac{1}{2}\right)^2 = p^2 + p + \frac{1}{4} < p^2 + p + 1 \leq b_p$$

the inequality

$$n^2 + (2p+1)\,n + b_p < \left(n + p + \frac{1}{2}\right)^2 \Leftrightarrow b_p < \left(p + \frac{1}{2}\right)^2$$

is not true anymore. The inequality in the left-hand side of (3) will hold for any $b_p \in \{p^2 + 1, ..., p^2 + 2p\}$ only if the inequality

$$2\left(p + \frac{2}{5}\right)n + \left(p + \frac{2}{5}\right)^2 \leq (2p+1)n + b_p,$$

equivalent to

$$\left(p + \frac{2}{5}\right)^2 - b_p \leq \frac{n}{5}, \tag{5}$$

holds, for any $n \in \mathbb{N}$ (or maybe starting from a certain rank). But this is always true, because (5) is verified for any $n \in \mathbb{N}, n \geq n_p$, where

$$n_p = \left[5\left(p+\frac{2}{5}\right)^2 - 5b_p\right] + 1.$$

Therefore $N_p$ will have the same first decimal digit, for every $n \geq 3$, only if (5) holds for $n = 3$, i.e. for those values of $p$ for which

$$b_p > p^2 + \frac{4}{5}p + \frac{19}{25}. \qquad \square$$

**Remark.** We can reformulate the previous problem more precisely as

**Problem 6e.** *Show that if* $b_p \in \{p^2+1, p^2+2, ..., p^2+p\}, p \in \mathbb{N}^*$, *then the first decimal of the number*

$$N_p = \sqrt{n^2 + (2p+1)n + b_p}$$

*is equal to* 4, *for any* $n \in \mathbb{N}, n \geq n_p$.

**Examples.** 1) In the case $p = 2$, when $b_p \in \{5, 6\}$ we obtain that the first decimal of the number $\sqrt{n^2+5n+5}$ is $a = 4$, for any $n \in \mathbb{N}$, and that the number $\sqrt{n^2+5n+6}$ has the same first decimal, $a = 4$, for any $n \in \mathbb{N}$;

2) In the case $p = 3$, when $b_p \in \{10, 11, ..., 15\}$, the condition $b_p < p^2 + p$ reduces the values to $b_p \in \{10, 11\}$ and $\sqrt{n^2+7n+10}$ has 4 as its first decimal only for $n \geq 10$, while $\sqrt{n^2+7n+11}$ has the same property for any $n \geq 3$.

## 6.4 Other Directions of Investigation

Once the first (constant) decimal of the irrational numbers in the previous section has been determined, we can go further and proceed in finding the next decimal. Thus we may obtain new problems of the following kind.

**Problem 6f.** *Find the first two decimals of the number*

$$\sqrt{n^2+11n+30}, n \in \mathbb{N}, n \geq 1.$$

**Solution.** We already know that

$$\sqrt{n^2+11n+30} = \overline{(n+5).4...}$$

and, further, the second decimal will be the first decimal of the number

$$10\left(\sqrt{n^2+11n+30} - n - 5\right) - 4.$$

For this we will follow the method shown when solving Problem 6.

We find in the end that the second decimal is 8 for $n \geq 1$. Therefore

$$\sqrt{n^2 + 11n + 30} = \overline{(n+5).48...}, \forall n \in \mathbb{N}^*. \qquad \square$$

Another direction of investigation, suggested by the problems 18689* from *Gazeta Matematica* no. 4/1981 and 19103* from the same journal, no. 2–3/1982, both proposed by L. Parsan, could reveal the existence of similar properties for other square roots (as in Problems 6–6f) as well as for numbers of the form

$$\sqrt[3]{n(n+1)(n+2)+b}, \ \sqrt[4]{n(n+1)(n+2)(n+3)+b}$$

or of the form

$$\left(\sqrt{n+\alpha} + \sqrt{n+\beta}\right)^2 .$$

We leave the readers the mission of exploring this, and end this section by presenting three other problems in the same class.

**Problem 6g.** *Show that the first decimal of the numbers*

$$\sqrt{n^2 + 5n + 5}\ ,\ \sqrt{n^2 + 5n + 6}$$

*is the same,* $\forall n \in \mathbb{N}$.

**Problem 6h.** *Determine* $b \in \mathbb{N}$ *so that the first decimal of the number* $\sqrt{n^2 + 7n + b}$ *is the same digit* $c$, $\forall n \in \mathbb{N}$. *Find this digit.*

**Problem 6i.** *Find the natural numbers* $a$ *and* $b$ *for which the integer part of the irrational number* $x = \sqrt{n^2 + an + b}$ *is* $n + 3$, *and its first decimal is a non-zero digit,* $c, \forall n \in \mathbb{N}^*, n \geq 3$. *Find* $c$.

**Solution.** We get $a = 7$, $b \in \{11, 12\}$ and $c = 4$. (The solution $a = 6$, $b = 9$, when we would get $x = n + 3 \in \mathbb{N}$ , can't be accepted). $\square$

## 6.5 Bibliographical Comments

Problem 6 appeared in *Gazeta Matematica* no. 1/1996 as Problem 23462, proposed by Irina Haivas.

# Theme 7

# Some Elementary Problems Regarding the Polynomial Approximation of Continuous Functions

In this section we will study how by using polynomials to approximate some functions which are continuous on a compact interval. The main part of our presentation here is contained in [1]. The investigations will have as a starting point the first problem presented there.

## 7.1 Source Problem

**Problem 7.** *Find all pairs* $(p, q)$ *for which the inequality*

$$\left|\sqrt{1-x^2} - px - q\right| \leq \frac{1}{2}\left(\sqrt{2}-1\right)$$

*holds, for all* $x \in [0,1]$.

**Solution.** An equivalent way of writing the given inequality is

$$\sqrt{1-x^2} - \frac{\sqrt{2}-1}{2} \leq px + q \leq \sqrt{1-x^2} + \frac{\sqrt{2}-1}{2}, \ \forall\, x \in [0,1]. \tag{1}$$

Let $g, h : [0,1) \to \mathbb{R}$ be given respectively by

$$g(h) = \sqrt{1-x^2} - \frac{\sqrt{2}-1}{2}, \quad x \in [0,1],$$

and

$$h(x) = \sqrt{1-x^2} + \frac{\sqrt{2}-1}{2}, \quad x \in [0,1].$$

We denote by $A, B$, respectively, the two extremities of the curve arc (see Figure 6)

$$y = g(x)\,,\; x \in [0,1]$$

and by $C, D$, respectively, the two extremities of the arc

$$y = h(x)\,, \quad x \in [0,1].$$

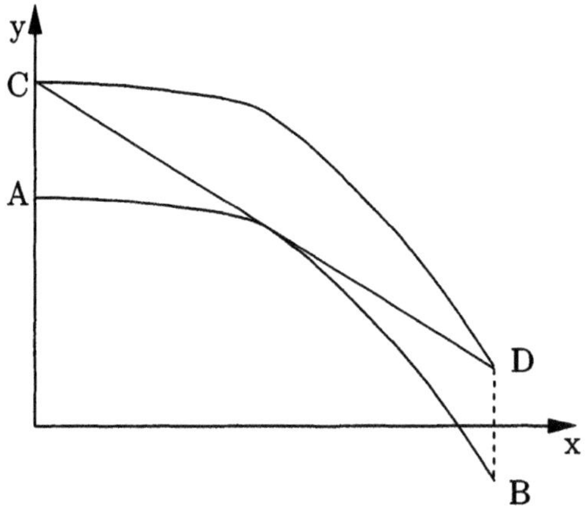


Figure 6

We have

$$A\left(0, \frac{3-\sqrt{2}}{2}\right); B\left(1, \frac{1-\sqrt{2}}{2}\right); C\left(0, \frac{1+\sqrt{2}}{2}\right) \text{ and } D\left(1, \frac{\sqrt{2}-1}{2}\right).$$

Studying their graphs or using any other method, we can state that $g$ and $h$ are concave functions.

The inequality (1) holds if and only if the line given by

$$y = px + q$$

is situated between the graphs of $g$ and $h$, for any $x \in [0,1]$.

We analyze now the line that contains the above considered points $C$ and $D$. This line is given by the equation

$$y = -x + \frac{\sqrt{2}+1}{2}$$

and, taking into account that $h$ is concave or directly by computing, we obtain that

$$-x + \frac{\sqrt{2}+1}{2} \leq h(x)\,,\; x \in [0,1],$$

which is equivalent to $1 - x \le \sqrt{1-x^2}$, $x \in [0,1]$, or if squared to the obvious inequality

$$-2x(1-x) \le 0\,, \ \forall\, x \in [0,1].$$

We also have that

$$g(x) \le -x + \frac{\sqrt{2}+1}{2}\,, \ \forall\, x \in [0,1], \tag{2}$$

i.e.,

$$\left(\sqrt{2}x - 1\right)^2 \ge 0\,, \ \forall\, x \in [0,1].$$

As for the equation attached to the inequality (2), i.e., $g(x) = -x + \dfrac{\sqrt{2}+1}{2}$, $x = \dfrac{1}{2}$ is a double root, we conclude that the line $CD$ is situated under the arc $CD$, above the arc $AB$ and is also tangent to the last one.

Hence the only line $y = px + q$ that satisfies the given conditions in Problem 7 is obtained for the pair

$$(p,q) = \left(-1, \frac{\sqrt{2}+1}{2}\right). \qquad \square$$

## 7.2 Exploring the Source Problem

Analyzing the solution given above for Problem 7 one can notice that, in order to get to the conclusion, one essential condition is the fact that the line we are looking for is a tangent to the graph of $g$ and besides it is situated between the graphs of the two functions $g$ and $h$. The unique approximating polynomial is in this case a first order polynomial. Instead of the $f(x) = \sqrt{1-x^2}$ function in Problem 7, one may well choose other functions, as in the examples below.

**Problem 7a.** *Show that there is a unique first order polynomial $P$ such that*

$$\left|8x^2 + P(x)\right| \le 1\,, \ \forall\, x \in [0,1].$$

**Problem 7b.** *Show that there is a unique polynomial $P(x) = ax + b$ such that*

$$\left|\sqrt{1-x} - P(x)\right| \le \frac{1}{8}\,, \ \forall\, x \in [0,1].$$

**Solution of Problem 7a.** The given inequality is equivalent to

$$\left|x^2 - Q(x)\right| \le \frac{1}{8}\,, \ \forall\, x \in [0,1], \tag{3}$$

where $Q(x) = -\frac{1}{8}P(x) = ax + b$, i.e.,

$$x^2 - \frac{1}{8} \le ax + b \le x^2 + \frac{1}{8}\,, \ \forall\, x \in [0,1].$$

We denote by $A, B$, respectively, the extremities of the parabola arc $y = x^2 - \frac{1}{8}$, $x \in [0,1]$ and by $C, D$, respectively, the extremities of the parabola arc $y = x^2 + \frac{1}{8}$, $x \in [0,1]$. So we have (see Figure 7):

$$A\left(0, -\frac{1}{8}\right);\ B\left(1, \frac{7}{8}\right);\ C\left(0, \frac{1}{8}\right) \text{ and } D\left(1, \frac{9}{8}\right).$$

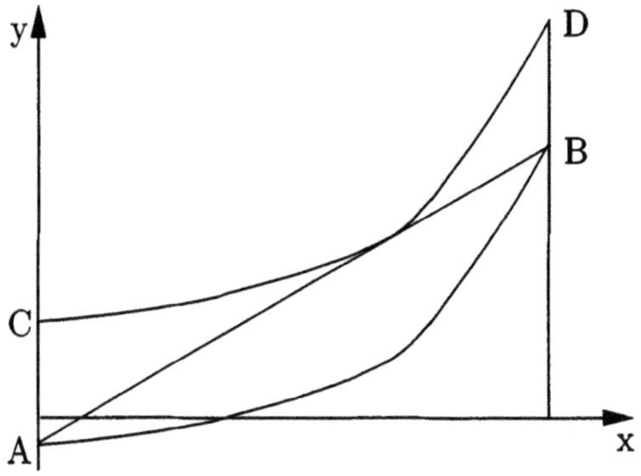


Figure 7

The polynomial $Q$ we are looking for will satisfy the given conditions only if the line

$$y = ax + b, \quad x \in [0,1]$$

is situated between the two parabola arcs $AB$ and $CD$.

We consider now the line $AB$, with the equation

$$y = x - \frac{1}{8},$$

a line that is situated above the parabola arc $AB$ — this is deduced from the inequality

$$x - \frac{1}{8} \geq x^2 - \frac{1}{8}, \quad x \in [0,1], \text{ i.e., } x(1-x) \geq 0, \quad x \in [0,1],$$

or from the fact that $y = x^2 - \dfrac{1}{8}$ is a convex parabola.

We also have that

$$x - \frac{1}{8} \leq x^2 + \frac{1}{8}, \quad x \in [0,1], \text{ i.e., } (2x-1)^2 \geq 0, \quad x \in [0,1],$$

and, as for the attached equation $x - \dfrac{1}{8} = x^2 + \dfrac{1}{8}$, we know that $x = \dfrac{1}{2}$ is a double root, we obtain that the line $AB$ is tangent to the parabola arc $CD$.

This proves that $AB$ is the only line situated between the parabola arcs $AB$ and $CD$, and so $Q(x) = x - \dfrac{1}{8}$ is the only first order polynomial that satisfies (3).

Therefore $P(x) = -8Q(x) = -8x + 1$ is the only first order polynomial to satisfy the given conditions in Problem 7a. □

**Solution of Problem 7b.** The given inequality is equivalent to

$$\sqrt{1-x} - \frac{1}{8} \leq ax + b \leq \sqrt{1-x} + \frac{1}{8}, \ \forall\, x \in [0,1]. \tag{4}$$

We are going to follow the same steps as above.

First we introduce the notations: $A, B$ for the extremities of the arc $y = \sqrt{1-x} - \frac{1}{8}$, $x \in [0,1]$ and $C, D$ for the extremities of the arc $y = \sqrt{1-x} + \frac{1}{8}$, $x \in [0,1]$.

So we have (see Figure 8):

$$A\left(0, \frac{7}{8}\right); B\left(1, -\frac{1}{8}\right); C\left(0, \frac{9}{8}\right) \text{ and } D\left(1, \frac{1}{8}\right).$$

It appears obvious now that the polynomial $P = ax + b$ will satisfy the given conditions only if the line segment $y = ax + b$, $x \in [0,1]$ is situated between the arcs $AB$ and $CD$.

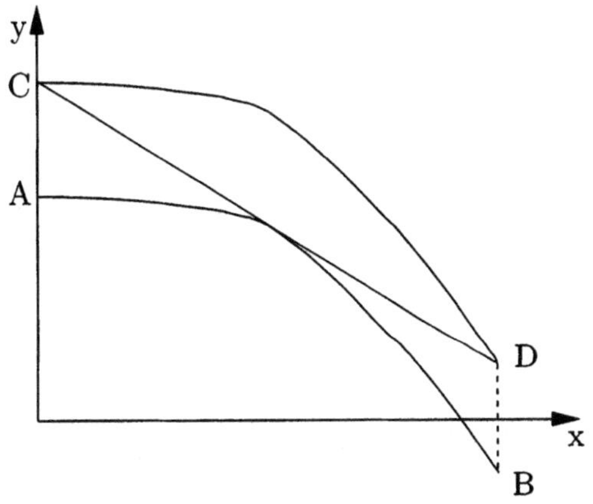


Figure 8

Now we consider the $CD$ line, which has the equation $y = -x + \frac{9}{8}$. This satisfies the condition (4), as

$$-x + \frac{9}{8} \leq \sqrt{1-x} + \frac{1}{8} \Leftrightarrow x(1-x) \geq 0,$$

which holds for any $x \in [0,1]$, and, respectively,

$$\sqrt{1-x} - \frac{1}{8} \leq -x + \frac{9}{8} \Leftrightarrow (4x-3)^2 \geq 0,$$

which is obviously true. Besides, $x = \frac{3}{4}$ is the unique solution of the equation

$$\sqrt{1-x} - \frac{1}{8} = -x + \frac{9}{8}.$$

All these prove that the line segment $CD$ is situated under the arc $CD$, above the arc $AB$ and is also tangent to the $AB$ arc. Therefore $CD$ is the only line having this property.

This leads to the fact that $P(x) = -x + \frac{9}{8}$ is the unique solution of the Problem 7b. □

**Remark.** We have solved Problems 7, 7a and 7b at a rather elementary level, that of the first high school class, thus avoiding an appeal to properties like continuity or concavity/convexity obtained by derivatives. That is why our solution lacks rigor in exploiting the conditions for a line to be tangent to an arc.

We can formulate a more general result now, a result that can be found in [2].

**Proposition 1.** *Let* $g, h : [0,1] \to \mathbb{R}$ *be two functions such that:*

*a)* $\forall x \in [0,1]$, $h(x) = (1-x) \cdot g(0) + x \cdot g(1)$;

*b)* $g$ *is continuous and* $g(x) \geq h(x)$, $\forall x \in [0,1]$.

*Then the function* $f : [0,1] \to \mathbb{R}$,

$$f(x) = h(x) + k, \ x \in [0,1], \tag{5}$$

*where*

$$k = \frac{1}{2} \cdot \max\{g(x) - h(x) \mid x \in [0,1]\}, \tag{6}$$

*is the unique linear function for which*

$$|g(x) - f(x)| \leq k, \ \forall x \in [0,1]. \tag{7}$$

*Proof.* From (6) we have that $g(x) - h(x) \leq 2k$, $\forall x \in [0,1]$, and from the condition b) we have that $0 \leq g(x) - h(x)$, $\forall x \in [0,1]$, which leads to

$$-k \leq g(x) - [h(x) + k] \leq k,$$

i.e.,

$$|g(x) - f(x)| \leq k, \ \forall x \in [0,1]$$

which shows that the function $f$ given by (5) satisfies condition (7).

We are going to prove that $f$ is the only function having this property. We assume that there is another linear function $u : \mathbb{R} \to \mathbb{R}$, different from $f$, that satisfies (7). We consider $k > 0$ (the case $k = 0$ is obvious).

As $g - h$ is continuous on the compact interval $[0,1]$, it is also bounded and it reaches its bounds. Hence there exists $x_0 \in [0,1]$ satisfying

$$(g-h)(x_0) = 2k.$$

We cannot have $x_0 = 0$ or $x_0 = 1$, as $(g-h)(0) = (g-h)(1) = 0 < 2k$ , and so $x_0 \in (0,1)$. As $u$ satisfies (7), we have that $u(x) \leq g(x) + k$, which leads to

$$u(0) \leq g(0) + k = h(0) + k = f(0).$$

In a similar way we obtain that $u(1) \leq f(1)$.

It is also easy to prove that for two given linear functions $u, f : \mathbb{R} \to \mathbb{R}$ satisfying $f(0) \geq u(0)$ and $f(1) \geq u(1)$, we have

$$f(x) > u(x)\,,\ \forall\, x \in (0,1). \tag{8}$$

But taking into account (7) and the property of $x_0$, we obtain

$$u(x_0) \geq g(x_0) - k = [h(x_0) + 2k] - k = h(x_0) + k = f(x_0),$$

which contradicts (8).

This shows that $f$ is unique. □

**Remark.** The proposition above can be easily adapted to the situation when b) is replaced by:

b′) $g$ is continuous and $g(x) \leq h(x)$, $\forall\, x \in [0,1]$,

a situation that implies some changes in formulating (5) and (6), too (for example, this is the case of convex functions).

**Particular Cases.** $1^0$. For $g(x) = \sqrt{1-x}$, when $h(x) = 1-x$, $\forall\, x \in [0,1]$, we obtain $k = \dfrac{1}{8}$, i.e., exactly Problem 7b;

$2^0$. For $g(x) = \sqrt{1-x^2}$, when $h(x) = 1-x, \forall x \in [0,1]$, we obtain $k = \dfrac{\sqrt{2}-1}{2}$, i.e., exactly Problem 7;

$3^0$. For $g(x) = -x^2$, when $h(x) = -x$ and $k = \dfrac{1}{8}$, we obtain Problem 7a.

At the end of this section we present some new problems that were obtained using the recipe given by Proposition 1. We also recommend the study of [1], where the problems belonging to this class are solved by means of two non-elementary methods.

**Problem 7c.** *Show that there is a unique polynomial $P(x) = ax + b$ such that*

$$\left|\frac{x}{x^2+1} - P(x)\right| \leq \frac{\sqrt{\sqrt{5}-2}}{2\left(\sqrt{5}+1\right)},\ \textit{for any } x \in [-1,1].$$

**Problem 7d.** *Find the polynomial $P = ax + b$ for which*

$$\left|\frac{1}{3-x} - P(x)\right| \leq \frac{1}{8\left(3+2\sqrt{2}\right)},\ \textit{for any } x \in [-1,1].$$

*How many such polynomials are there?*

**Problem 7e.** *Find a first order polynomial $P(x)$ for which*

$$\left|\frac{x}{1-x} - P(x)\right| \le \frac{1}{6+4\sqrt{2}}, \text{ for any } x \in \left[0, \frac{1}{2}\right],$$

*and prove that it is the only one with this property.*

## 7.3 Bibliographical Comments

The source problem of this section is the fourth problem given at the fourth test (May 30, 1983) for selecting the Romanian team for the $24^{th}$ IMO, Paris, 1983, and was published in *Gazeta Matematica* no. 10/1983, p. 394.

Problem 7a was proposed by the author of this book at the National Mathematics Olympiad, the county round, Maramures, in 1993, while Problem 7b was published in *Gazeta Matematica* no. 9/1994 as problem 23070, under the signature of the same author. This problem was solved by the editor Al. Constantinescu, in *Gazeta Matematica* no. 6/1995, pp. 276–277. The solution is basically the one given by the author and there is also mentioned that "it lacks some rigor, although from the geometrical point of view it is clear. So, we ask the readers for a complete solution and a general study upon this theme". The one who answered this invitation was D. Serbanescu, by the note [2], published in the centennial issue of *Gazeta Matematica* no. 9/1995.

Problem 7c is a stronger variant of the problem C:23 which appeared in *Gazeta Matematica* no. 3/1980, proposed by L. Panaitopol. Originally the constant in the right-hand side was $0.1 > \dfrac{\sqrt{\sqrt{5}-2}}{2\left(\sqrt{5}+1\right)}$, a situation when the uniqueness of the first order polynomial is not assured.

Problem 7d starts from problem 18968 in *Gazeta Matematica* no. 10/1981, having A. Atanasiu as author, but there one has to find a third order polynomial.

Problem 7e corresponds to problem 18538 from *Gazeta Matematica* no. 7/1980, written by M. Chirita and having a statement similar to that of the previous problem.

## 7.4 References


1. **S.G. Gal, V. Berinde,** *Elementary Problems in Polynomial Approximation of Continuous Functions.* (Romanian), Lucr. Semin. Creativ. Mat., **3** (1993-1994), 1–22

2. **D. Serbanescu,** *On the Problem 23070.* (Romanian), Gazeta Matematica, **100** (1995), no. 9, 508-509.

# Theme 8

# On an Interesting Divisibility Problem

## 8.1 Source Problem

**Problem 8.** *Find the natural number $N$ so that $N \cdot 8888$ contains all the digits of the decimal system, each taken once.*

**Solution.** We notice that the problem can be reformulated as follows: find a natural number, consisting of 10 distinct digits and divisible by 8888. We will now solve the problem in this last formulation. Let

$$n = \overline{a_1 a_2 \ldots a_{10}} \tag{1}$$

be the required number.

Because $8888 = 2^3 \cdot 11 \cdot 101$, further we will need to apply criteria of divisibility by 8, 11 and 101. These criteria will be deduced below, as the old textbooks for forms V and VI ([2] and [3]) don't include any of them.

**Proposition 1.** *Let $n$ be a natural number in the form (1). Then*

1) *$n$ is divisible by 8 if and only if the number*

$$N_1 = 4a_8 + 2a_9 + a_{10}$$

*is divisible by 8;*

2) *$n$ is divisible by 11 if and only if the number*

$$N_2 = a_1 + a_3 + a_5 + a_7 + a_9 - (a_2 + a_4 + a_6 + a_8 + a_{10})$$

*is divisible by 11;*

3) *$n$ is divisible by 101 if and only if the number*

$$N_3 = 10(a_1 - a_3 + a_5 - a_7 + a_9) + a_2 - a_4 + a_6 - a_8 + a_{10}$$

*is divisible by 101.*

*Proof.* 1) We write $n$ in the following way:

$$\begin{aligned} n = \;& 1000 \cdot \overline{a_1a_2\ldots a_7} + a_8 \cdot 10^2 + a_9 \cdot 10 + a_{10} = 8 \cdot 125 \cdot \overline{a_1a_2\ldots a_7} \\ & +(96+4) \cdot a_8 + (8+2) \cdot a_9 + a_{10} = M8 + 4a_8 + 2a_9 + a_{10}. \end{aligned}$$

2) Here we use the fact that

$$10 = 11 - 1 \text{ and } 10^2 = 9 \cdot 11 + 1$$

and then $n$ can be written as:

$$\begin{aligned} n &= 10^2 \cdot \overline{a_1a_2...a_8} + a_9 \cdot 10 + a_{10} = M_{11} + \overline{a_1a_2...a_8} - a_9 + a_{10} \\ &= M_{11} + 10^2 \cdot \overline{a_1a_2...a_6} + 10 \cdot a_7 + a_8 - a_9 + a_{10} \\ &= M_{11} + \overline{a_1a_2...a_6} - a_7 + a_8 - a_9 + a_{10} \\ &= \ldots = M_{11} + a_1 - a_2 + a_3 - a_4 + \ldots - a_9 + a_{10}, \end{aligned}$$

which proves 2).

3) As $10^2 = 101 - 1$ we will write

$$\begin{aligned} n &= 10^2 \cdot \overline{a_1a_2...a_8} + a_9 \cdot 10 + a_{10} \\ &= (101-1) \cdot \overline{a_1a_2...a_8} + 10a_9 + a_{10} = M_{101} - \overline{a_1a_2...a_8} + 10a_9 + a_{10} \\ &= M_{101} - 10^2 \cdot \overline{a_1a_2...a_6} - 10a_7 - a_8 + 10a_9 + a_{10} \\ &= \ldots = M_{101} + 10(a_1 - a_3 + a_5 - a_7 + a_9) + a_2 - a_4 + a_6 - a_8 + a_{10}, \end{aligned}$$

thus getting the criterion at 3). □

**Remark**. It is easy to notice that the first statement in Proposition 1 is true for any natural number of $k$ digits, while 2) and 3) can be adjusted to the general case without any difficulty.

Coming back to the solution of Problem 8, let us denote

$$\begin{aligned} n_1 &= a_1 + a_3 + a_5 + a_7 + a_9, \\ n_2 &= a_2 + a_4 + a_6 + a_8 + a_{10}, \\ n_3 &= a_1 - a_3 + a_5 - a_7 + a_9, \\ n_4 &= a_2 - a_4 + a_6 - a_8 + a_{10}, \end{aligned}$$

and $N_1, N_2$ and $N_3$ be as in Proposition 1. Because, in terms of the problem, the digits $a_1, a_2, \ldots, a_{10}$ belong to the set $\{0, 1, 2, \ldots, 9\}$ and are distinct, we must have

$$a_1 + a_2 + \ldots + a_{10} = n_1 + n_2 = 45(= 0 + 1 + 2 + \ldots + 9).$$

Therefore, the number $N_2 = n_1 - n_2$ is necessarily odd and, according to Proposition 1, $N_2$ must be divisible by 11. We deduce that

$$N_2 \in \{-11, 11\}.$$

If $N_2 = -11$, i.e., $n_2 - n_1 = 11$, as $n_1 + n_2 = 45$, we find $n_1 = 17$ and $n_2 = 28$ while for $N_2 = 11$ i.e., $n_1 - n_2 = 11$, we determine $n_1 = 28$, $n_2 = 17$. We will tackle the first case, with $n_1 = 17$ and $n_2 = 28$.

We have to write the set of all digits in the decimal number system as the union of two disjoint sets, say $M_1$ and $M_2$, each having five elements and such that the sum of the elements in $M_1$ is 17 and of those in $M_2$ is 28. Among all the possible splittings we consider

$$M_1 = \{1, 2, 3, 5, 6\}, \quad M_2 = \{0, 4, 7, 8, 9\}. \tag{2}$$

(We see that there are many such splittings. Here there are two of them

$$\begin{aligned} M_1 &= \{0, 2, 4, 5, 6\}, \quad M_2 = \{1, 3, 7, 8, 9\}, \\ M_1 &= \{1, 2, 3, 4, 7\}, \quad M_2 = \{0, 5, 6, 8, 9\}.) \end{aligned}$$

We return to the splitting (2), for which we will analyze the divisibility by 101. Because

$$n_3 = n_1 - 2(a_3 + a_7) \text{ and } n_4 = n_2 - 2(a_4 + a_8), \tag{3}$$

on the one hand, and $n_1, n_2 \in \{17, 28\}$, on the other hand, we deduce we must have $-17 \le n_3, n_4 \le 26$. Therefore

$$N_3 \le 26 \cdot 10 + 26,$$

which shows that the acceptable values for $N_3$ are given by

$$N_3 \in \{-101, 101, 202\}.$$

On the other hand, the relations (3) show that the pairs $n_1$ and $n_3$, and $n_2$ and $n_4$, respectively, have the same parity. Then, if for example we have $N_3 = 101$, then the expansion $101 = 10 \cdot n_3 + n_4$ implies that $n_4$ is odd, hence $n_2 = 17$ and then $n_1 = 28$ and therefore $n_3$ is even.

From the same relation we obtain that $n_3 = 10$ and $n_4 = 1$,

$$a_3 + a_7 = \frac{28 - 10}{2} = 9 \text{ and } a_4 + a_8 = \frac{17 - 1}{2} = 8.$$

Because $n_1 = 28$ and $n_2 = 17$, we will look for $a_1, a_3, \ldots, a_9$ in the set $M_2$ and for $a_2, a_4, \ldots, a_{10}$ in the set $M_1$. From the splitting we have chosen we get

$$a_3, a_7 \in \{0, 9\} \text{ and } a_4, a_8 \in \{3, 5\}.$$

Thus, $a_9 \in \{4, 7, 8\}$ and $a_{10} \in \{2, 6\}$, because $n \,\vdots\, 888$ implies that $a_{10}$ has to be even. Applying the criterion of divisibility by 8, we obtain that the convenient digits are $a_8 = 3$, $a_9 = 7$ and $a_{10} = 6$, from which we get

$$a_4 = 5, \; a_2, a_6 \in \{1, 2\} \text{ and } a_1, a_5 \in \{4, 8\}.$$

Taking $a_3 = 0$, $a_7 = 9$, $a_2 = 1$, $a_6 = 2$, $a_1 = 4$ and $a_5 = 8$ we get an answer for the problem. The required number is

$$n = 4105829376$$

for which we have

$$4105829376 = 8888 \cdot 461952$$

and so the required number $N$ in Problem 8 is 461952. □

## 8.2 Other Related Problems

The solution of Problem 8 shows that this one doesn't have a unique answer and the solution method can be easily applied to other problems of the same type. We can obtain other problems and we will present some of them here.

**Problem 8a.** *Find a natural number $n$ so that $n \cdot 9999$ is written with all the digits of the decimal system, each used only once.*

**Problem 8b.** *Find the least natural number $n$ for which $n \cdot 7777$ is written with all the digits of the decimal system, each used only once.*

**Problem 8c.** *Find the greatest natural number of 10 distinct digits, divisible by 6666.*

**Problem 8d.** *How many natural numbers of 10 distinct digits and divisible by 5555 do exist?*

**Problem 8e.** *Determine the least and the greatest number, respectively, formed of 10 distinct digits and divisible by 1999.*

**Answer.** The least number is

$$1023759864 = 512136 \times 1999$$

and the greatest one is

$$9874630215 = 4939785 \times 1999.$$

(The solutions were obtained using the computer, see [4].) □

**Problem 8f.** *The same statement as in Problem 8e, but using 1995 instead of 1999.*

**Answer.** ([4]) The least number is

$$1023476895 = 513021 \times 1995$$

and the greatest one is

$$9876351240 = 4950552 \times 1995.$$

**Problem 8g.** *Find the least and respectively the greatest square written using all the digits of the decimal system, each taken only once.*

**Answer.** ([4]) The numbers are, respectively,

$$1026753849 = 32043^2,$$
$$9814072356 = 99066^2.$$

We invite readers to discover by themselves other properties of these numbers.

Over this whole section we have worked only in the decimal number system, but it is obvious that one can formulate similar problems using other number systems.

## 8.3 Bibliographical Comments

We don't know the original source of Problem 8. We learned it from a radio program on *Radio Romania "Tineret"*, in the 1980s.

Problem 8a was proposed by the author and appeared in *Gazeta Matematica* no. 4/1994 as problem O.G.:160 but, unfortunately, the statement was affected by a printing error and so instead of $n \cdot 9999$ it had $n \cdot 1999$. This error was corrected later in no. 3/1994 through an Errata of the author but meanwhile the problem had been solved using the computer by T. Zvonaru and the results were reported in the note [4]. Anyway, this story showed that even a printing error can lead to the discovery of new problems, as in the note quoted above. Problems 8c–8g are taken after [4].

## 8.4 References


1. **V. Berinde,** *An Interesting Divisibility Problem.* (Romanian), Lucr. Semin. Creativ. Mat., **4** (1993-1994), 1–6

2. **C.P. Popovici et al.,** *Mathematics. Textbook for Form V.* (Romanian), Editura Didactica si Pedagogica, 1986

3. **C.P. Popoviciu, I.C. Ligor,** *Mathematics. Arithmetics. Algebra. Textbook for Form VI.* (Romanian), Editura Didactica si Pedagogica, 1985

4. **T. Zvonaru,** *On the Problem O.G:160 from Gazeta Matematica no. 4/1993.* (Romanian), Gazeta Matematica, **99** (1994), no. 11, 503–505.

# Theme 9

# Determinants with Alternate Entries

Starting from a problem regarding the computation of the determinant of a square matrix having alternate entries, the present section illustrates how we can enrich this class of problems by exploring and investigating around their essential aspects.

## 9.1 Source Problem

**Problem 9.** *Let* $A, B \in \mathcal{M}_n(\mathbb{R})$, $A = (a_{ij})$, $B = (b_{ij})$ *be two square matrices given respectively by*

$$a_{ij} = (-1)^{\max(i,j)},\ b_{ij} = (-1)^{\min(i,j)}, i, j = 1, 2, ..., n.$$

*Show that*

$$\det A = \det B.$$

**Solution.** Let us denote $d_n = \det A$ and $d_n' = \det B$. We have

$$d_n = \begin{vmatrix} -1 & 1 & -1 & 1 & \dots & (-1)^n \\ 1 & 1 & -1 & 1 & \dots & (-1)^n \\ -1 & -1 & -1 & 1 & \dots & (-1)^n \\ \cdot & \cdot & \cdot & \cdot & \cdot & \cdot \\ (-1)^n & (-1)^n & (-1)^n & (-1)^n & \dots & (-1)^n \end{vmatrix}$$

and by means of the transformation column 1− column 2→ column 1 (abbreviated $C_1 - C_2 \to C_1$) we get

$$d_n = \begin{vmatrix} -2 & 1 & -1 & 1 & \dots & (-1)^n \\ 0 & 1 & -1 & 1 & \dots & (-1)^n \\ 0 & -1 & -1 & 1 & \dots & (-1)^n \\ . & . & . & . & \dots & . \\ 0 & (-1)^n & (-1)^n & (-1)^n & \dots & (-1)^n \end{vmatrix} \tag{1}$$
$$= (-2)\begin{vmatrix} 1 & -1 & 1 & \dots & (-1)^n \\ -1 & -1 & 1 & \dots & (-1)^n \\ 1 & 1 & 1 & \dots & (-1)^n \\ . & . & . & \dots & . \\ (-1)^n & (-1)^n & (-1)^n & \dots & (-1)^n \end{vmatrix}$$

(we have expanded the determinant over the first column).

The order of the determinant in the right-hand side of (1) is $n-1$. By factoring (-1) on each row of this last determinant we obtain from (1) that

$$d_n = (-2)\cdot(-1)^{n-1}\cdot\begin{vmatrix} -1 & 1 & -1 & \dots & (-1)^{n-1} \\ 1 & 1 & -1 & \dots & (-1)^{n-1} \\ . & . & . & \dots & . \\ (-1)^{n-1} & (-1)^{n-1} & (-1)^{n-1} & \dots & (-1)^{n-1} \end{vmatrix} \tag{2}$$

and the determinant in the right-hand side of (2) is just $d_{n-1}$. Therefore, we deduce that $(d_n)$ satisfies

$$d_n = 2\cdot(-1)^n\cdot d_{n-1},\quad n \geq 3,$$

which gives, after straightforward calculations,

$$d_n = 2^{n-1}\cdot(-1)^{\frac{n^2+n+4}{2}},\quad n \geq 2, \tag{3}$$

since $d_2 = \begin{vmatrix} -1 & 1 \\ 1 & 1 \end{vmatrix} = -2.$

Now, in order to compute $d'_n$, we add its second row to the first one ($r_1+r_2 \to r_1$) and then expand the obtained determinant. It results that

$$d'_n = \begin{vmatrix} 1 & 1 & 1 & \dots & 1 \\ 1 & -1 & -1 & \dots & -1 \\ . & . & . & \dots & . \\ 1 & -1 & 1 & \dots & (-1)^n \end{vmatrix},$$

and now, by factoring $(-1)$ on each row, it results that $(d'_n)$ satisfies a recurrence relation similar to (3):

$$d'_n = 2\cdot(-1)^n\cdot d'_{n-1}.$$

Using the fact that $d'_2 = -2$, we finally get $d_n = d'_n,\ \forall\, n \geq 2.$

## 9.2 A Close Examination of the Problem and its Solution. Directions of Investigation

Let us notice that the functions $\max(i,j)$ and $\min(i,j)$, respectively, have induced partially or totally alternate signs in the entries of the rows and columns of matrices $A$ and $B$. We exploited these alternate elements by adding (or subtracting) columns or rows, in order to get many null entries. Therefore, we may tackle in a similar way the computation of the determinants of matrices defined by means of other alternative functions. Let us begin with the function $(-1)^{i+j}$. Thus we obtain

**Problem 9a.** *Compute* $\det A$, *where* $A = (a_{ij})$, $a_{ij} = (-1)^{i+j}$, *for all* $i,j \in \{1,2,\ldots,n\}$, $n \geq 2$.

**Solution.** Let us denote $d_n = \det A$. We get

$$d_n = \begin{vmatrix} 1 & -1 & 1 & -1 & \ldots & (-1)^{n+1} \\ -1 & 1 & -1 & 1 & \ldots & (-1)^{n+2} \\ . & . & . & . & \ldots & . \\ (-1)^{n+1} & (-1)^{n+2} & (-1)^{n+3} & (-1)^{n+4} & \ldots & (-1)^{n+n} \end{vmatrix}.$$

If we add the second row (column) to the first row (column) we shall obtain a null row (column). Hence $d_n = 0\,,\ \ \forall\, n \geq 2$. □

**Remark.** In order to avoid the trivial case of Problem 9a, we can define in different ways the upper- and sub-diagonal parts of the matrix. Thus we obtain new interesting problems, which are nicer and much more difficult.

**Problem 9b.** *Compute* $\det A$, $A = (a_{ij})_{1 \leq i,j \leq n}$, *where*

$$a_{ij} = \begin{cases} (-1)^{i+j} \cdot a, & \text{if } i < j, \\ \quad x \quad, & \text{if } i = j, \\ (-1)^{i+j} \cdot b, & \text{if } i > j. \end{cases}$$

**Solution.** We use the *linearity* of the determinants, a property that can be expressed in the following way: if the entries on the column $k$ are all of the form

$$a_{ik} = a'_{ik} + a''_{ik}\,, \quad i = 1,2,\ldots,n,$$

then $\det A$ may be written as the sum of two determinants, which have on the column $k$ the entries $a'_{ik}$ and $a''_{ik}$, respectively, the other entries being identical to that in $\det A$. (The same property occurs for rows as well). We can write

$$d_n = \begin{vmatrix} x & -a & a & -a & \ldots \\ -b & x & -a & a & \ldots \\ b & -b & x & -a & \ldots \\ -b & b & -b & x & \ldots \\ . & . & . & . & \ldots \end{vmatrix} = \begin{vmatrix} (x-a) & +a & -a & a & \ldots \\ 0 & -b & x & -a & \ldots \\ 0 & +b & -b & x & \ldots \\ 0 & -b & b & -b & \ldots \\ . & . & . & . & \ldots \end{vmatrix}$$

and by applying the linearity property on column 1 we get

$$d_n = \begin{vmatrix} x-a & -a & a & \dots \\ 0 & x & -a & \dots \\ 0 & -b & x & \dots \\ 0 & b & -b & \dots \\ . & . & . & \dots \end{vmatrix} + \begin{vmatrix} a & -a & a & \dots \\ -b & x & -a & \dots \\ b & -b & x & \dots \\ -b & b & -b & \dots \\ . & . & . & \dots \end{vmatrix} . \tag{4}$$

The first determinant in the right-hand side of (4) may be expanded on the first column to obtain $(x-a) \cdot d_{n-1}$, while the second one, denoted by $d_n'$, will be computed in the following way: we factor $a$ on the first row and then apply the transformations

$$r_2 + b \cdot r_1 \rightarrow r_2 \,;\; r_3 - b \cdot r_1 \rightarrow r_3 \,;\; r_4 + b \cdot r_1 \rightarrow r_4, \dots .$$

We get

$$d_n' = \begin{vmatrix} 1 & -1 & 1 & \dots \\ 0 & x-b & b-a & \dots \\ 0 & 0 & x-b & \dots \\ 0 & 0 & 0 & x-b \\ . & . & . & \dots \end{vmatrix} ,$$

a determinant whose sub-diagonal entries are all null, and hence its value is the product of all entries in the main diagonal

$$d_n' = a \cdot (x-b)^{n-1},$$

and so $(d_n)$ will satisfy the recurrence relation

$$d_n = (x-a) \cdot d_{n-1} + a \cdot (x-b)^{n-1}, \quad n \geq 2. \tag{5}$$

Now, as the numbers $a, b$ and $x$ are arbitrarily taken, we have to discuss the following situations.

**Case 1.** If $x = a$, i.e., all the diagonal entries are equal to $a$ and we have alternative signs in the sub-diagonal part, then

$$d_n = a \cdot (a-b)^{n-1}\,, \quad n \geq 1.$$

In particular, for $a = b = 1$, we obtain Problem 9a.

**Case 2.** If $x \neq a$, then we shall write (5) in the equivalent form

$$\frac{d_n}{(x-a)^n} = \frac{d_{n-1}}{(x-a)^{n-1}} + a \cdot \frac{(x-b)^{n-1}}{(x-a)^n},$$

and sum it for $n$ in $\{2, 3, \dots, n\}$ to obtain, after all cancellations and computations, that

$$d_n = (x-a)^{n-1} \cdot d_1 + a(x-a)^{n-1} \cdot \sum_{k=1}^{n-1} \left(\frac{x-b}{x-a}\right)^k, \; n \geq 2. \tag{6}$$

In order to compute the sum, say $S$, in the right-hand side of (6) we shall analyze the sub-cases:

**2a)** If $a = b$, that is $\dfrac{x-b}{x-a} = 1$, then $S = n-1$ and using $d_1 = x$ it results that

$$d_n = (x-a)^{n-1}\left[x + (n-1)\cdot a\right], \quad n \geq 2.$$

**2b)** If $a \neq b$, then $S$ is a geometric progression of ratio $\dfrac{x-b}{x-a} \neq 1$. We finally get

$$d_n = x(x-a)^{n-1} + \frac{a(x-b)}{b-a}\left[(x-a)^{n-1} - (x-b)^{n-1}\right], \quad n \geq 2. \qquad \square$$

Some particular cases of Problem 9b are interesting by themselves and have been proposed to different mathematics competitions.

**Problem 9c.** *Compute* $\det A$, *where* $A = (a_{ij})_{1\leq ij\leq n}$, *and*

$$a_{ij} = \begin{cases} (-1)^{i+j}\cdot a, & \text{if } i < j, \\ 1, & \text{if } i = j, \\ (-1)^{i+j}\cdot b, & \text{if } i > j. \end{cases}$$

**Problem 9d.** *The same statement, with* $a = 3$ *and* $b = 2$.

**Remark.** We can obtain other related problems by considering other functions $f(i,j)$ which possess the alternative sign property, as for instance

$$f(i,j) = (-1)^{\left[\frac{i+j}{2}\right]},$$

where $[x]$ denotes the integer part of the real number $x$.

Another example is shown by the next problem.

**Problem 9e.** *Compute* $\det A$, *where* $A = (a_{ij}) \in \mathcal{M}_n(\mathbb{R})$ *and*

$$a_{ij} = \begin{cases} (-1)^{|i-j|}, & \text{if } i \neq j, \\ a, & \text{if } i = j. \end{cases}$$

## 9.3 Other Directions of Investigation

We notice that all the problems already considered in this section have the same particularity: the entries, excepting those in the main diagonal, alternate their sign. Let us try to see what happens when all the entries in the upper or sub-diagonal part of the matrix are identical (and of the same sign, of course). We start by solving a simple problem of this kind.

**Problem 9f.** *Let* $A = (a_{ij}) \in \mathcal{M}_n(\mathbb{R})$,

$$a_{ij} = \begin{cases} a, & \text{if } i \leq j, \\ (-1)^{i+j}\cdot b, & \text{if } i > j, \end{cases} \quad i,j \in \{1,2,\ldots,n\}.$$

*Compute* $\det A$.

**Solution.** This problem is not at all similar to Problem 9a. Let us denote $d_n = \det A$. Then

$$d_n = \begin{vmatrix} a & a & a & \dots & a \\ -b & a & a & \dots & a \\ b & -b & a & \dots & a \\ . & . & . & \dots & . \end{vmatrix}.$$

Obviously if $a = 0$, then $d_n = 0$, $\forall a \geq 1$.

After applying the transformations

$$C_1 := C_1 - C_n \ , \ C_2 := C_2 - C_n \ , \dots, \ C_{n-1} := C_{n-1} - C_n$$

we obtain

$$d_n = \begin{vmatrix} 0 & 0 & 0 & \dots & a \\ -b-a & 0 & 0 & \dots & a \\ b-a & -b-a & 0 & \dots & a \\ -b-a & b-a & -b-a & \dots & a \\ . & . & . & \dots & a \end{vmatrix}.$$

Then we expand the last determinant over the first row and notice that the new determinant of order $n-1$ obtained in this way has null upper-diagonal entries. It finally results that

$$d_n = (-1)^{n+1} \cdot a \cdot (-b-a)^{n-1} = a(a+b)^{n-1}, \ n \geq 2. \qquad \square$$

**Remark.** When $b = -a$ we get $\det A = 0$.

Problem 9a suggests that we consider a more general statement given by

**Problem 9g.** *Compute* $\det A$, *where* $A = (a_{ij}) \in \mathcal{M}_n(\mathbb{R})$ *and*

$$a_{ij} = \begin{cases} a & , \ if \ \ i > j, \\ x & , \ if \ \ i = j, \\ (-1)^{i+j} \cdot b, & if \ \ i < j. \end{cases}$$

**Hint.** We proceed in a similar way as for solving Problem 9a. Let us denote $d_n = \det A$. Then

$$d_n = \begin{vmatrix} x & -b & b & -b & \dots \\ a & x & -b & b & \dots \\ a & a & x & -b & \dots \\ a & a & a & x & \dots \\ . & . & . & . & \dots \end{vmatrix} = \begin{vmatrix} (x-a)+a & -b & b & -b & \dots \\ 0+a & x & -b & b & \dots \\ 0+a & a & x & -b & \dots \\ 0+a & a & a & x & \dots \\ . & . & . & . & \dots \end{vmatrix}$$

and so, using the linearity property of the determinants, it results that

$$d_n = (x-a) \cdot d_{n-1} + d'_n \ ,$$

with $d'_n$ a determinant that can be easily computed only for $x = a$. $\square$

We recommend that readers take on themselves the task of computing $d'_n$ (and $d_n$ as well) in the general case and also suggest that they try to solve the next similar (but generally more difficult) problems.

**Problem 9h.** *Compute* $\det A$, *where* $A = (a_{ij}) \in \mathcal{M}_n(\mathbb{R})$ *and*

$$a_{ij} = \begin{cases} 1 & , \text{ if } i \geq j, \\ (-1)^{i+j} \cdot j, & \text{if } i < j. \end{cases}$$

**Problem 9i.** *Idem for*

$$a_{ij} = \begin{cases} a & , \text{ if } i > j, \\ b & , \text{ if } i = j, \\ (-1)^{i+j} \cdot j, & \text{if } i < j. \end{cases}$$

**Problem 9j.** *Idem for*

$$a_{ij} = \begin{cases} (-1)^{i+j} \cdot a^j, & \text{if } i < j, \\ b & , \text{ if } i \geq j. \end{cases}$$

**Problem 9k.** *Idem for*

$$a_{ij} = \begin{cases} (-1)^{i+j} \cdot j, & \text{if } i < j, \\ x & , \text{ if } i = j, \\ (-1)^{i+j} \cdot i, & \text{if } i > j. \end{cases}$$

**Problem 9l.** *Idem for*

$$a_{ij} = \sin(i+j)\frac{\pi}{2} \quad , \ i, j \in \{1, 2, ..., n\}.$$

**Problem 9m.** *Idem for*

$$a_{ij} = \begin{cases} \alpha_j^{i-1} & , \text{ if } i > j, \\ (-1)^{i+j} \cdot \alpha_j^{i-1}, & \text{if } i \leq j, \end{cases}$$

*(alternate Vandermonde determinant).*

## 9.4 Bibliographical Comments

Problem 9 was published in *Gazeta Matematica* no. 12/1981 as problem 19035*, being proposed by Marius Dadarlat. Problem 9b was published in *Revista de Matematica a Elevilor din Timisoara* no. 2/1986 as problem 6000, proposed by V. Berinde.

Problem 9c, having the same author, was proposed to the *Romanian National Mathematical Olympiad*, the county round (Maramures), in 1993, while problem 9d was proposed at the same competition in 1996.

Excepting for Problems 9h and 9i, the other Problems in this class (9g, 9j–9m) seem to be more difficult and the author has not had enough time (so far) to solve them completely. This is the reason we warmly recommend to the readers to do this work, without forgetting to look for other combinations of alternative functions, in order to obtain any other related problems.

# Theme 10

# Solving Some Cyclic Systems

A system of the form

$$\begin{cases} f_1(x_1, x_2, x_3, \ldots, x_n) = 0, \\ f_2(x_1, x_2, x_3, \ldots, x_n) = 0, \\ \ldots\ldots\ldots\ldots\ldots\ldots\ldots\ldots \\ f_n(x_1, x_2, x_3, \ldots, x_n) = 0 \end{cases} \tag{*}$$

is called **cyclic** if, by any circular permutation of its unknowns

$$x_1, x_2, \ldots, x_n,$$

the system does not change. The symmetric systems of two equations and two unknowns are the most simple examples of cyclic systems.

The present section presents a unitary algorithm for solving a class of cyclic systems based on the fixed point principle and then shows how this kind of problems could be enriched by means of a creative approach.

By introducing a fixed point technique to solve cyclic systems we do not want to limit the reader to this method, since there are many other heuristic methods to solve them, see for instance those indicated in Section 4, where we discuss some other cyclic systems.

## 10.1 Source Problem

**Problem 10.** *Solve the cyclic system*

$$2x = y + \frac{2}{y}; \; 2y = z + \frac{2}{z}; \;\; 2z = x + \frac{2}{x}.$$

## 10.2 Solving the Source Problem

**Solution 1.** We must have $x \neq 0,\ y \neq 0,\ z \neq 0$. Let us consider for the beginning $x > 0$. From the third equation it results that $z > 0$, and then by the second equation we find $y > 0$. Therefore if $(x, y, z)$ is a solution, then either $x > 0,\ y > 0,\ z > 0$ or $x < 0,\ y < 0,\ z < 0$.

We consider the first case. If we take $a := \sqrt{y}$ and $b := \sqrt{2/y}$ in the obvious inequality $a^2 + b^2 \geq 2ab$ (valid for any $a, b \in \mathbb{R}$), then we get $y + \frac{2}{y} \geq 2\sqrt{2}$ which shows, based on the first equation, that $x \geq \sqrt{2}$. Since the system is cyclic, it also results that $y \geq \sqrt{2}$ and $z \geq \sqrt{2}$. If we now sum all the equations, side by side, we get

$$x + y + z = 2\left(\frac{1}{x} + \frac{1}{y} + \frac{1}{z}\right) \tag{1}$$

and as $x \geq \sqrt{2} \Leftrightarrow \frac{1}{x} \leq \frac{1}{\sqrt{2}}$, and other similar relations for $y$ and $z$ hold as well, from (1) it results that

$$3\sqrt{2} \leq x + y + z = 2\left(\frac{1}{x} + \frac{1}{y} + \frac{1}{z}\right) \leq 2 \cdot 3 \cdot \frac{1}{\sqrt{2}} = 3\sqrt{2}$$

which shows that in the previous inequalities we actually have equality. Hence

$$x + y + z = 3\sqrt{2} \Longleftrightarrow \left(x - \sqrt{2}\right) + \left(y - \sqrt{2}\right) + \left(z - \sqrt{2}\right) = 0,$$

and, based on $x - \sqrt{2} \geq 0,\ y - \sqrt{2} \geq 0,\ z - \sqrt{2} \geq 0$, this implies $x = y = z = \sqrt{2}$. Therefore $(\sqrt{2}, \sqrt{2}, \sqrt{2})$ is the unique positive solution of the system.

If $x < 0$ (when $y < 0,\ z < 0$ as well) let us denote $x' = -x,\ x > 0$ etc. and apply the above result. It results that $(-\sqrt{2}, -\sqrt{2}, -\sqrt{2})$ is the unique negative solution. Therefore, our system has two solutions

$$(\sqrt{2}, \sqrt{2}, \sqrt{2}) \text{ and } (-\sqrt{2}, -\sqrt{2}, -\sqrt{2}). \qquad \square$$

**Solution 2.** Let us restrict to the case $x > 0,\ y > 0,\ z > 0$. From the first equation we get

$$y^2 - 2xy + 2 = 0,$$

which has real roots, since $x \geq \sqrt{2}$ implies $\Delta_y = x^2 - 2 \geq 0$. Only the root $y = x + \sqrt{x^2 - 2}$ meets the conditions in the problem, since $x > \sqrt{2}$ implies $x - \sqrt{x^2 - 2} < \sqrt{2}$. Therefore $y = x + \sqrt{x^2 - 2}$ and, similarly, $z = y + \sqrt{y^2 - 2}$, $x = z + \sqrt{z^2 - 2}$. By summing these equations we get

$$\sqrt{x^2 - 2} + \sqrt{y^2 - 2} + \sqrt{z^2 - 2} = 0$$

which necessarily implies $x = y = z = \sqrt{2}$. The second part follows as in Solution 1. $\square$

**Remark.** If we try to solve the system by the substitution method, we are led to a very complicated equation

$$2x = \frac{\frac{x}{2}+\frac{1}{x}}{2} + \frac{1}{\frac{x}{2}+\frac{1}{x}} + \frac{2}{\frac{\frac{x}{2}+\frac{1}{x}}{2} + \frac{1}{\frac{x}{2}+\frac{1}{x}}} \tag{2}$$

which discourages even the most tenacious problem solvers. Then should we completely abandon this idea? Not at all !

We shall see that, if we rewrite equation (2) in a convenient form, we are in the position to find a very important and useful method for solving this kind of problems, as shown below.

**Solution 3.** Let $f : [\sqrt{2}, \infty) \to \mathbb{R},\ f(x) = \frac{x}{2} + \frac{1}{x}$. The system can be written as

$$x = f(y),\ y = f(z),\ z = f(x)$$

which shows that equation (2) is actually

$$x = f^{\,3}(x), \tag{2'}$$

where we denote $f^3 = f \circ f \circ f$ (from Solution 1 we know that $f(x) \geq 2$ and so the composition $\circ$ is well defined). From the remark above, it is now very clear for us that equation (2′) cannot be solved directly. We consider instead of (2′) the much simpler equation

$$x = f(x). \tag{3}$$

It is easy to show that if $x = a$ is a solution of (3), then it is a solution of (2′), too. But the reverse is not generally true. In the next paragraph we shall give sufficient conditions for the equivalence of the two equations (2′) and (3). We shall see that, in our particular case, the equation (2′) has the unique positive solution $x = \sqrt{2}$ which is, in turn, the unique positive solution of (3). On the other hand, if $a$ is a solution of (2′), then $(a, a, a)$ is a solution of the system

$$x = f(y),\ y = f(z),\ z = f(x)$$

and conversely, any solution of the previous system is of the form $(a, a, a)$, with $a$ a solution of the equation (2′), due to the fact that the system is cyclic.

In our case, equation (3) will be $x = \frac{x}{2} + \frac{1}{x} \iff x^2 = 2$ which has the unique positive solution $a = \sqrt{2}$. Hence $(\sqrt{2}, \sqrt{2}, \sqrt{2})$ is the positive solution of the system. The negative solution can be obtained in a similar way as by the previous two solutions. □

More details regarding this method are given in the next section.

## 10.3 A Unitary Method for Solving a Class of Cyclic Systems

If the system (*) can be equivalently written in the form

$$x_1 = f(x_2),\ x_2 = f(x_3), \ldots, x_{n-1} = f(x_n), x_n = f(x_1), \tag{4}$$

then by substituting $x_2, x_3, \ldots, x_n$ with respect to $x_1$ we deduce that $x_1$ is a solution of the equation

$$x = \underbrace{(f \circ f \circ \ldots \circ f)}_{n \text{ times}}(x) \iff x = f^n(x). \tag{5}$$

The same property holds for $x_2, x_3, \ldots, x_n$, too, which shows that if $a$ is a solution of equation (5), then $(a, a, \ldots, a)$ is a solution of the system (4) and conversely, any solution of the system (4) is of the form $(a, a, \ldots, a)$ with $a$ a solution of (5).

Therefore, in order to find the solutions of the system (4), we have to solve the equation (5). The last one is however not an easy task, even for $n = 3$, as it could be seen in the previous paragraph. We will show in the following that, in certain cases, equation (5) is equivalent to the equation

$$x = f(x) \tag{6}$$

which is in fact a very simple fixed point problem.

We give a brief survey on fixed point theory on the real axis (for the more general case of the metric spaces see for example [6]).

**Definition 1.** Let $A \subset \mathbb{R}$ be a nonempty set and let $f : A \to \mathbb{R}$ be a given function. A number $a \in A$ is called a **fixed point of** $f$ if $f(a) = a$.

The set of all fixed points of $f$ is denoted by $Fix\ f$ or $F_f$ :

$$F_f = \{x \in A \mid f(x) = x\}.$$

**Example 1. a)** The function $f : \mathbb{R} \to \mathbb{R}$, $f(x) = x^2 + 5x + 4$ has a unique fixed point, $F_f = \{-2\}$;

**b)** The function $f : \mathbb{R} \to \mathbb{R}$, $f(x) = x^2 - x$ has two fixed points, $F_f = \{0, 2\}$;

**c)** The function $f : \mathbb{R} \to \mathbb{R}$, $f(x) = x + 2$ has no fixed points ($F_f = \emptyset$), while the function $g : \mathbb{R} \to \mathbb{R}$, $g(x) = x$ has an infinite set of fixed points, $F_g = \mathbb{R}$.

**Geometric Interpretation.** In the case of numerical functions, which are considered in the present section, $F_f$ represents the set of points on the real axis that correspond to the intersection points of the graph of $f$ with the first bisectrix (see Figure 9), since the equation (6) is equivalent to the system $\begin{cases} y = f(x), \\ y = x. \end{cases}$

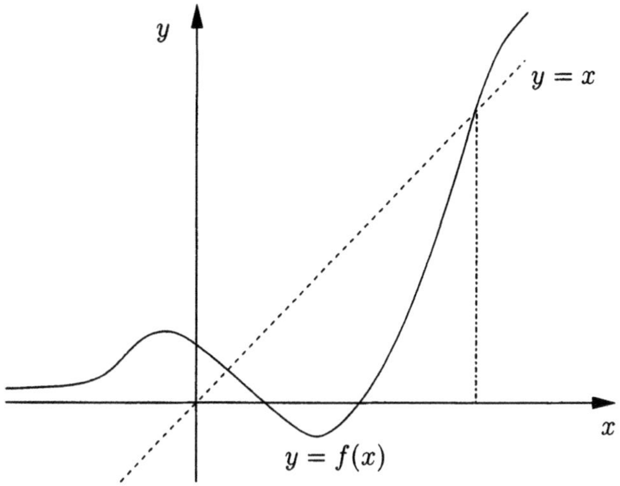


Figure 9

Let $f : A \to \mathbb{R}$ be a given function and $(f^n)_{n\geq 1}$be the sequence of its iterates, that is, $f^1 = f$ and $f^n = f \circ f^{n-1}$, for all $n \geq 2$.

**Proposition 1.** *If* $a \in F_f$, *then* $a \in F_{f^n}$, $\forall\, n \in \mathbb{N}^*$.

**Proof.** We have $a \in F_f \Leftrightarrow f(a) = a \Rightarrow f(f(a)) = f(a) = a$, hence $f^2(a) = a$ and so on, $f^n(a) = a$, $\forall\, n \in \mathbb{N}^*$, $n \geq 2$. □

**Remark.** Proposition 1 expresses in terms of the fixed point theory the fact that any solution of equation (6) is a solution of (5), too. A natural question then arises: is the converse of Proposition 1 also true? The following result offers a partial answer to this question.

**Proposition 2.** *If* $F_{f^n} = \{a\}$ *for all* $n \geq 2$, *then* $F_f = \{a\}$.

**Proof.** The condition $F_{f^n} = \{a\}$ shows that $a$ is the unique fixed point of each iterate $f^n$, $n \geq 2$. Since

$$f^{n+1}(a) = f^n(\, f\,(\,a\,)\,)$$

and, on the other hand,

$$f^{n+1}(\, f\,(a)\,) = f\,(\, f^n(a)\,) = f\,(a),$$

we obtain that

$$f^n\,(\, f\,(a)\,) = f(a), \text{ that is } f(a) \in F_{f^n},$$

and since $F_{f^n} = \{a\}$, we get $f(a) = a$, i.e., $a \in F_f$.

The function $f$ cannot have other fixed points. Indeed, if $b \in F_f$, then by Proposition 2 we have that $b \in F_{f^n}$, i.e., $b = a$. Therefore $F_f = \{a\}$. □

**Remark.** The condition that $f^n$ has a unique fixed point is essential, as shown by

**Example 2.** The function $f : \{1,2,3\} \to \{1,2,3\}$ with $f(1) = 3$, $f(2) = 2$ and $f(3) = 1$ has $F_f = \{2\}$ and $F_{f^2} = \{1,2,3\}$.

We are interested in finding sufficient conditions to ensure that the set $F_{f^n}$ has a unique element. In this case, in view of Proposition 2, we should have

$$F_{f^n} = F_f, \text{ for all } n \geq 2.$$

**Proposition 3.** *Let* $f : A \to A$ *be increasing. If* $F_f = \{a\}$, *then* $F_{f^n} = \{a\}$, *for any* $n \in \mathbb{N}^*$.

**Proof.** Let $b \in A$, $b \neq a$. We will prove that $b \notin F_{f^n}$. Since $F_f = \{a\}$, it results that $f(b) \neq b$, say $f(b) > b$ (the case $f(b) < b$ will be similarly tackled). Since $f$ is increasing, we have that

$$f(f(b)) \geq f(b) > b, \text{ i.e., } f^2(b) > b,$$

and so on,

$$f^n(b) > b, \ \forall n \in \mathbb{N}^*,$$

and hence $b \notin F_{f^n}$. □

**Remark.** Proposition 3 is not true anymore if $f$ is a decreasing function, as shown by the next example.

**Example 3.** The function $f : \mathbb{R} \to \mathbb{R}$, $f(x) = -x$, $x \in \mathbb{R}$ is decreasing but $\{0\} = F_f \neq F_{f^2} = \mathbb{R}$.

**Theorem 1.** (The Contraction Mapping Principle) *Let* $I$ *be a closed interval in* $\mathbb{R}$ *and* $f : I \to I$ *be a contraction, that is, a function satisfying*

$$|f(x) - f(y)| \leq c \cdot |x - y|, \ \forall x, y \in I, \tag{7}$$

*where* $c$ *is a constant,* $0 \leq c < 1$. *Then there exists a unique* $a \in I$ *such that*

$$F_f = F_{f^n} = \{a\}, \ \forall n \in \mathbb{N}^*.$$

The proof of Theorem 1 requires some knowledge of calculus and we will not present it here. For the interested reader we refer to [3] and [6].

**Remark.** The number $c$ in condition (7), called the **contraction coefficient (constant)**, does not depend on $x$ and $y$.

Based on these fixed point principles, we are now able to design a unitary method for solving the cyclic systems in the form (4), which will consist in the following steps:

**Step 1.** We write down the system in the form (4) (thus identifying the function $f$);
**Step 2.** We find the fixed points of $f$ (i.e., the set $F_f$);
**Step 3.** We search the subsets $A$ such that we could apply Proposition 3 or Theorem 1 in order to obtain $\{a\} = F_f = F_{f^n}$;
**Step 4.** For any $a$ obtained at Step 3, $(a, a, \ldots, a)$ is a solution for the cyclic system.

**Applications.** We are now ready to solve some old and new cyclic systems by means of the above described method.

**Problem 10a.** *Solve in* $\mathbb{R}$ *the system*

$$\begin{cases} x = \sqrt{y+45} - \sqrt{y+5}\,, \\ y = \sqrt{z+45} - \sqrt{z+5}\,, \\ z = \sqrt{x+45} - \sqrt{x+5}\,. \end{cases}$$

**Solution 1.** If we add all the equations and denote

$$E(x) = x + \sqrt{x+5} - \sqrt{x+45},$$

we get

$$E(x) + E(y) + E(z) = 0. \tag{8}$$

But

$$\begin{aligned} E(x) &= x - 4 + \sqrt{x+5} - 3 - \left(\sqrt{x+45} - 7\right) \\ &= (x-4)\left[1 + \frac{1}{3+\sqrt{x+5}} - \frac{1}{7+\sqrt{x+45}}\right] \end{aligned}$$

and since

$$3 + \sqrt{x+5} < 7 + \sqrt{x+45},\ \forall x \geq -5,$$

we have $signum E(x) = signum(x-4)$. This shows that in both cases

$$a)\ x, y, z \in [2, \infty) \quad \text{or} \quad b)\ x, y, z \in [-5, 2]$$

the equation (8), and hence the system itself, has the unique solution

$$x = y = z = 2.$$

□

**Solution 2.** (Using the fixed point principle)
**Step 1.** The system may be written as

$$x = f(y),\ y = f(z),\ z = f(x)$$

with $f : [-5, \infty) \to \mathbb{R}$, $f(x) = \sqrt{x+45} - \sqrt{x+5}$.
**Step 2.** The fixed points of $f$ are obtained by solving the equation

$$x = \sqrt{x+45} - \sqrt{x+5} \Leftrightarrow x = \frac{40}{\sqrt{x+45} + \sqrt{x+5}}.$$

In the left-hand side of the previous equation we have an increasing function, while in the right-hand side we have a decreasing function (prove it !). As $x = 4$ is a solution of the equation, it necessarily is the unique one, hence $F_f = \{4\}$.
**Step 3.** The function $f$ is strictly decreasing, so we cannot apply Proposition 3. Then we try to apply Theorem 1. To this end, let us first notice that for any $x$ in $[-5, \infty)$ we have $f(x) \geq 0$.

We therefore consider the restriction of $f$ to the interval $I = [0, \infty)$, which will also be denoted by $f$. We have $f(I) \subset I$ and

$$\begin{aligned} |f(x) - f(y)| &= \left|\sqrt{x+45} - \sqrt{y+45} - \left(\sqrt{x+5} - \sqrt{y+5}\right)\right| \\ &= \left|\frac{x-y}{\sqrt{x+45}+\sqrt{y+45}} - \frac{x-y}{\sqrt{x+5}+\sqrt{y+5}}\right| = |x-y| \cdot |G(x,y)|, \end{aligned}$$

where

$$\begin{aligned} |G(x,y)| &= \left|\frac{1}{\sqrt{x+45}+\sqrt{y+45}} - \frac{1}{\sqrt{x+5}+\sqrt{y+5}}\right| \\ &\leq \frac{1}{\sqrt{x+45}+\sqrt{y+45}} + \frac{1}{\sqrt{x+5}+\sqrt{y+5}} \leq \frac{1}{2\sqrt{45}} + \frac{1}{2\sqrt{5}}, \quad \forall\, x, y \in I, \end{aligned}$$

which shows that

$$|f(x) - f(y)| \leq c \cdot |x - y|, \ \forall\, x, y \in I, \text{ with } c = \frac{1}{2\sqrt{45}} + \frac{1}{2\sqrt{5}} < 1.$$

All the conditions in Theorem 1 are satisfied and hence:
**Step 4.** The unique solution of the system is $(4, 4, 4)$. □

**Remark.** It is easy to notice that our method can be applied regardless of the number of equations (and unknowns) in the system.

The next problem illustrates the case where $F_f$ contains more than one element.

**Problem 10b.** *Solve the system*

$$\begin{cases} 2x^3 - 7x^2 + 8x - 2 = y, \\ 2y^3 - 7y^2 + 8y - 2 = z, \\ 2z^3 - 7z^2 + 8z - 2 = x. \end{cases}$$

**Solution. Step 1.** We have $f(x) = 2x^3 - 7x^2 + 8x - 2$ and hence
**Step 2.** $F_f = \left\{\frac{1}{2}, 1, 2\right\}$.

**Step 3.** We first consider the restriction of $f$ to the interval $I_1 = \left(-\infty, \frac{1}{2}\right]$. Since

$$f'(x) = 6x^2 - 14x + 8 = 2(3x-4)(x-1),$$

we have that $f'(x) > 0$, $\forall\, x \in I_1$ and hence $f$ is increasing on $I_1$. Therefore, for any $x \in I_1$ $\left(\Leftrightarrow x \leq \frac{1}{2}\right)$, we have $f(x) \leq f\left(\frac{1}{2}\right) = \frac{1}{2}$ which shows that $f(I_1) \subset I_1$ and we can apply Proposition 3.

**Step 4.** $(\frac{1}{2}, \frac{1}{2}, \frac{1}{2})$ is the unique solution of the system in $I_1$.

In a similar manner we obtain the other two solutions $(1, 1, 1)$ and $(2, 2, 2)$ by considering $I_2 = (\frac{1}{2}, 1]$ and $I_3 = [2, \infty)$, respectively. □

## 10.4 Other Exploratory Directions

Using the same method we can easily solve the following systems.

**Problem 10c.** *Solve the system*

$$\begin{cases} 3^x + 4^y = 5^x, \\ 3^y + 4^z = 5^y, \\ 3^z + 4^x = 5^z. \end{cases}$$

**Problem 10d.** *Solve the system*

$$3^x + 4^x = 5^y;\ 3^y + 4^y = 5^z;\ 3^z + 4^z = 5^x.$$

**Problem 10e.** *Solve the system*

$$2^{x-1} + 2^{\frac{1}{\sqrt{y}}} = 3;\ 2^{y-1} + 2^{\frac{1}{\sqrt{z}}} = 3;\ 2^{z-1} + 2^{\frac{1}{\sqrt{x}}} = 3.$$

**Hint.** The fixed point problem reduces in this case to the equation

$$2^{x-1} + 2^{\frac{1}{\sqrt{x}}} = 3$$

which can be solved as follows. Since

$$\begin{aligned} 2^{x-1} + 2^{\frac{1}{\sqrt{x}}} &= \frac{2^x}{2} + 2^{\frac{1}{\sqrt{x}}} = \frac{2^x + 2 \cdot 2^{\frac{1}{\sqrt{x}}}}{2} \\ &= \frac{2^x + 2^{\frac{1}{\sqrt{x}}} + 2^{\frac{1}{\sqrt{x}}}}{2} \geq \frac{3}{2} \sqrt[3]{2^{x+\frac{2}{\sqrt{x}}}} \end{aligned}$$

and

$$x + 2/\sqrt{x} \geq 3,\ \forall\, x > 0$$

which is equivalent to

$$\left(\sqrt{x} - 1\right)^2 \left(\sqrt{x} + 2\right) \geq 0,$$

it results that $2^{x-1} + 2^{\frac{1}{\sqrt{x}}} \geq 3,\ \forall\, x > 0$, with equality if and only if $x = 1$.

Hence $(1, 1, 1)$ is the unique solution of the system. □

We leave to the readers the task of discovering other related problems and also of solving as many as possible of the 35 cyclic systems included in [3]. It is important to stress the limits of our method: except for systems in the form (4), it cannot be applied to all cyclic systems, as for example to the next systems.

**Problem 10f.** *Solve the system*

$$2x = y + \frac{2}{z};\ 2y = z + \frac{2}{x};\ 2z = x + \frac{2}{y}.$$

**Problem 10g.** *Solve in $\mathbb{R}$ the system*

$$\begin{cases} x = \sqrt{y+23} - \sqrt{z+7}, \\ y = \sqrt{z+23} - \sqrt{x+7}, \\ z = \sqrt{x+23} - \sqrt{y+7}. \end{cases}$$

**Problem 10h.** *Solve the system*

$$\begin{cases} 3^x + 4^y = 5^z, \\ 3^y + 4^z = 5^x, \\ 3^z + 4^x = 5^y. \end{cases}$$

And the list could be continued.

We notice that Problem 10h, which is proposed in [4], can be solved also by using the method presented in Section 4 of this book.

## 10.5 Bibliographical Comments

Problem 10 was proposed to a university entrance exam in the former Soviet Union (see [10], where its relationship to recurrence sequences is illustrated) and then in Romania (in 1981). We preferred it as a source problem even if it is not the oldest one in this class, especially for the fact that it can be solved by different methods which could be thus compared.

Problem 10a is a variant of problem 117 [1], while Problem 10b was published by *Matematika v skole* in 1985.

Problem 10c was proposed by the author of the book at the Intercounty Mathematics Competition "Gr.C. Moisil" in 1993, see [2], while Problem 10d was published in *Gazeta Matematica* no. 12/1996, as problem C:1878; it was also presented in Section 4 of this book (Problem 4r), where we can also find Problem 10c as Problem 4t.

Problems 10f and 10g are published here for the first time, while Problem 10h is contained in the list in [4].

In the paper [8] one can find some supplementary criteria to eliminate false solutions of cyclic systems that have been obtained by fixed point arguments.

## 10.6 References

1. **Gh. Andrei et al.**, *Algebra. Problems for the Mathematics Competitions.* (Romanian), Constanta, 1990

2. **D. Andrica, V. Berinde et al.**, *The Intercounty Mathematics Competition "Gr. C. Moisil". Editions I-XV (1986-2000).* (Romanian), Editura Cub Press 22, 2000

3. **V. Berinde,** *The Fixed Point Principle Applied for Solving Cyclic Systems.* (Romanian), Lucr. Semin. Creativ. Mat., **2** (1992-1993), 1–23

4. **V. Berinde,** *Two Solutions of a Competition Problem.* Mathematics Competitions, **11** (1998), no. 1, 36–44

5. **M. Ganga,** *Themes and Problems of Mathematics.* (Romanian), Editura Tehnica, 1990

6. **I.A. Rus,** *Principles and Applications of Fixed Point Theory.* (Romanian), Editura Dacia, 1979

7. **M. Ganga,** *Systems of Equations for High Schools* (Romanian), Editura Mathpress, 1996

8. **C. Caragea, R. Nicolescu,** *An Unitary Framework for Solving Systems of Equations.* (Romanian), Gazeta Matematica, **87** (1982), no. 12, 435–436

9. **M. Chirita, M. Piticari,** *A Method for Solving Systems of Equations in the Complex Numbers Set.* (Romanian), Gazeta Matematica, **100** (1995), no. 12, 708–711

10. **S. Cremarenco, B. Enescu,** *Using the Computer in Studying Some Recurrent Sequences.* (Romanian), Gazeta Matematica, **93** (1988), no. 11–12, 434–437

# Theme 11

# On a Property of Recurrent Affine Sequences

In the present section we shall give a characterization of convergent sequences of real numbers, by means of affine combinations of $k$ successive terms ($k \geq 2$). In this way we shall survey some known results and add other similar related problems.

The method presented here will allow readers, if the creative investigation will be continued, to obtain other interesting related results.

## 11.1 Source Problem

**Problem 11.** *If* $(a_n)$ *is a sequence of real numbers satisfying*

$$\lim_{n \to \infty} \left( a_{n+1} - \frac{1}{2} a_n \right) = 0,$$

*show that*

$$\lim_{n \to \infty} a_n = 0.$$

## 11.2 Solution of the Source Problem

We denote

$$x_n = a_{n+1} - \frac{1}{2} a_n, \ n \geq 0, \tag{1}$$

a relation that can be written in the form

$$2a_{n+1} = a_n + 2x_n \Leftrightarrow 2^{n+1} a_{n+1} = 2^n a_n + 2^{n+1} x_n.$$

We sum the previous equalities for $n \in \{0, 1, ..., n-1\}$ to obtain

$$2^n a_n = a_0 + \sum_{k=1}^{n} 2^k x_{k-1},$$

that is

$$a_n = \frac{1}{2^n}\left(a_0 + \sum_{k=1}^{n} 2^k x_{k-1}\right), \; n \geq 1. \tag{2}$$

Using now (2) we can show in several ways that the convergence of the sequence $(x_n)$ to zero implies the convergence of the sequence $(a_n)$ to zero.

The simplest one seems to be that based on the Cesaro–Stolz lemma (for other methods — more or less complicated — we refer to the papers [4],[5],[7]). Indeed, from (2) we have

$$\lim_{n\to\infty} a_n = \lim_{n\to\infty} \frac{a_0 + \sum\limits_{k=1}^{n} 2^k x_{k-1}}{2^n} = \lim_{n\to\infty} \frac{2^{n+1} \cdot x_n}{2^{n+1} - 2^n} = 2 \lim_{n\to\infty} x_n \tag{3}$$

which completes the solution. □

## 11.3 A Close Examination of the Problem and its Solution. Exploratory Directions

The solution of Problem 11 in the previous paragraph is very general, so it can be applied, basically in the same way, to solving more general problems. Here are four examples of this kind.

**Problem 11a.** *Let $(a_n)$ be a sequence of real numbers. Show that*

$$\lim_{n\to\infty} a_n = 0 \Leftrightarrow \lim_{n\to\infty}\left(a_{n+1} - \frac{1}{p} a_n\right) = 0\,, \;\; p \geq 2.$$

**Problem 11b.** *Prove that*

$$\lim_{n\to\infty} a_n = 0 \Leftrightarrow \lim_{n\to\infty} (a_{n+1} - b a_n) = 0\,, \;\; b \in (0,1).$$

**Problem 11c.** *Let $(b_n)$ be a sequence of real numbers such that $b_n \in (-1, 0), \forall\, n \geq 1$ and $\lim\limits_{n\to\infty} b_n = b \in (-1, 0]$ exists. Prove that the sequence $(x_n)$ converges if and only if the sequence $(x_{n+1} + b_n x_n)_{n\geq 1}$ converges.*

**Problem 11d.** *If the sequence $(x_n)_{n\geq 1}$ satisfies*

$$\lim_{n\to\infty} (x_{n+1} - \lambda x_n) = 0\,, \;\; \lambda \in (-1, 1),$$

*then it converges to zero.*

**Remarks.** 1) Let us mention that Problem 11a extends Problem 11 (which is obtained for $p = 2$), Problem 11b extends Problem 11a, while Problem 11c extends in turn Problem 11b, replacing the constant $b$ by a convergent sequence $(b_n)$;

2) We can reformulate Problem 11 by conveniently writing (3) as

$$\lim_{n\to\infty} a_n = \lim_{n\to\infty} (2a_{n+1} - a_n),$$

which can be in turn rewritten in the form

$$\lim_{n\to\infty} a_n = a \Leftrightarrow \lim_{n\to\infty} (2a_{n+1} - a_n) = a \tag{4}$$

that provides a characterization of convergent sequences by means of a linear (affine) combination of two of its successive terms.

A natural question then arises: could we give similar characterizations by means of affine combinations of the $k$ successive terms, with $k \geq 3$?

By following this direction of investigation we shall obtain new interesting problems. To this end we need the following result, actually a reformulation of Problem 11d for the sequences converging to a limit not necessarily zero.

**Problem 11e.** *If* $\lambda \in (-1, 1)$, *then*

$$\lim_{n\to\infty} a_n = a \Leftrightarrow \lim_{n\to\infty} (a_{n+1} - \lambda a_n) = (1 - \lambda) \cdot a.$$

**Solution.** If we denote, similarly to Problem 11,

$$x_n = a_{n+1} - \lambda a_n\,, \quad n \geq 0,$$

then we obtain in the same way

$$a_n = \lambda^n \left( a_0 + \sum_{k=1}^{n} \frac{x_{k-1}}{\lambda^k} \right), \quad n \geq 1.$$

**Case 1:** $\lambda \in (0, 1)$. We have $\lim\limits_{n\to\infty} \left(\dfrac{1}{\lambda}\right)^n = \infty$ and by the Stolz–Cesaro lemma we get

$$\lim_{n\to\infty} a_n = \lim_{n\to\infty} \frac{a_0 + \sum\limits_{k=1}^{n} \lambda^{-k} x_{k-1}}{\lambda^{-n}} = \lim_{n\to\infty} \frac{\lambda^{-n-1} x}{\lambda^{-n-1} - \lambda^{-n}} = \frac{1}{1-\lambda} \lim_{x\to\infty} x_n.$$

**Case 2:** $\lambda \in (-1, 0)$. In this situation $\left(\dfrac{1}{\lambda}\right)^n$ has two limit points $-\infty$ and $+\infty$ and we shall analyze separately the even and odd subsequences, respectively, obtaining the same result.

## 11.4 Recurrences with Many Successive Terms

For the beginning we consider recurrences with **three** successive terms. A typical problem for this case is

**Problem 11f.** *Let $(a_n)$ be a sequence of real numbers. Show that*

$$\lim_{n\to\infty} a_n = a \Leftrightarrow \lim_{n\to\infty}(4a_{n+2} - 4a_{n+1} + a_n) = a. \tag{5}$$

**Solution 1.** (Using Problem 11e) Since

$$\lim(4a_{n+2} - 4a_{n+1} + a_n) = a \Leftrightarrow \lim\left(a_{n+2} - a_{n+1} + \frac{1}{4}a_n\right) = \frac{1}{4}a$$

and

$$a_{n+2} - a_{n+1} + \frac{1}{4}a_n = a_{n+2} - \frac{1}{2}a_{n+1} - \frac{1}{2}\left(a_{n+1} - \frac{1}{2}a_n\right),$$

by denoting $y_n = a_{n+1} - \frac{1}{2}a_n$ the equality in the right-hand side of (5) may be written as

$$\lim_{n\to\infty}\left(y_{n+1} - \frac{1}{2}y_n\right) = \frac{1}{4}a \Leftrightarrow \lim_{n\to\infty}\left(y_{n+1} - \frac{1}{2}y_n\right) = \left(1 - \frac{1}{2}\right)\cdot\frac{a}{2}.$$

Using now Problem 11e (with $\lambda = \frac{1}{2}$) we know that

$$\lim_{n\to\infty}\left(y_{n+1} - \frac{1}{2}y_n\right) = \left(1 - \frac{1}{2}\right)\cdot\frac{a}{2} \Leftrightarrow \lim_{n\to\infty} y_n = \frac{a}{2},$$

that is

$$\lim_{n\to\infty}\left(a_{n+1} - \frac{1}{2}a_n\right) = \left(1 - \frac{1}{2}\right)\cdot a,$$

which is equivalent, considering Problem 11e, to $\lim_{n\to\infty} a_n = a$. □

**Solution 2.** (This is the way in which the author of this problem actually discovered Problem 11e). We consider the sequence $(b_n)_{n\geq 1}$, defined by

$$b_1 = a_1 \quad\text{and}\quad b_n = a_n - \frac{n+1}{2n}a_{n-1}, \text{ for } n \geq 2. \tag{6}$$

We rewrite the second relation in (6) in the equivalent form

$$a_n = \frac{n+1}{2n}\cdot a_{n-1} + b_n, \quad n \geq 2 \tag{6'}$$

and apply the same technique as in paragraph 11.2. This time we divide (6′) by $\frac{n+1}{2^n}$ and sum the obtained relations for $n \in \{2, 3, ..., n-1\}$ to get a result similar to (2), that is

$$a_n = \frac{(n+1)S_n}{2^n}, \; n \geq 2,$$

where $S_n = \sum\limits_{k=1}^{n} \dfrac{2^k b_k}{k+1}$ (we considered $a_0 = 0$).

We now compute $\lim\limits_{n\to\infty} a_n$ by applying twice the Stolz–Cesaro lemma: if the limit $\lim\limits_{n\to\infty} \dfrac{\alpha_{n+1} - \alpha_n}{\beta_{n+1} - \beta_n} = l$ exists, then $\lim\limits_{n\to\infty} a_n = \lim\limits_{n\to\infty} \dfrac{\alpha_n}{\beta_n}$ also exists and equals $l$. Let us denote $\alpha_n = (n+1)S_n$ and $\beta = 2^n$. After all the computations we get

$$\lim_{n\to\infty} \frac{\alpha_{n+1} - \alpha_n}{\beta_{n+1} - \beta_n} = \lim_{n\to\infty} \frac{S_n + 2^{n+1} b_{n+1}}{2^n}. \tag{7}$$

Then we apply again the Stolz–Cesaro lemma for computing the limit in the right-hand side of (7).

By denoting $\alpha'_n = S_n + 2^{n+1} b_{n+1}$ and $\beta'_n = 2^n$ we get

$$\frac{\alpha'_{n+1} - \alpha'_n}{\beta'_{n+1} - \beta'_n} = 2\left(2b_{n+2} - \frac{n+1}{n+2} b_{n+1}\right)$$

and by (6)

$$\frac{\alpha'_{n+1} - \alpha'_n}{\beta'_{n+1} - \beta'_n} = 4a_{n+2} - 4a_{n+1} + a_n.$$

Hence from $\lim\limits_{n\to\infty} (4a_{n+2} - 4a_{n+1} + a_n) = a$ it results that $\lim\limits_{n\to\infty} \dfrac{\alpha'_n}{\beta'_n} = \lim\limits_{n\to\infty} a_n = a$, which ends the proof (the "$\Rightarrow$" part is immediate). □

**Remark.** We notice that the second solution given to Problem 11f may be applied to the more general recurrence relation:

$$b_n = a_n - \frac{n+1}{cn} \cdot a_{n-1} \tag{6''}$$

obtained from (6′) by replacing 2 with a certain constant $c$. Thus we get

**Problem 11g.** *Show that for any real number $\lambda$, $|\lambda| > 1$,*

$$\lim_{n\to\infty} a_n = a \Leftrightarrow \lim_{n\to\infty} \left(\lambda^2 a_{n+2} - 2\lambda a_{n+1} + a_n\right) = (\lambda - 1)^2 \cdot a.$$

**Remark.** For $\lambda > 1$, the solution reproduces the case $\lambda = 2$ previously treated, while the case $\lambda < -1$ is easily reduced to the first one. Thus we obtain new problems of the following type.

**Problem 11h.** *Let $(x_n)_{n\geq 0}$ be a sequence of real numbers such that*

$$\lim_{n\to\infty} (4x_{n+2} + 4x_{n+1} + x_n) = 9.$$

*Show that* $\lim\limits_{n\to\infty} x_n = 1$.

**Answer.** We have $\lambda = -2$ and $a = 1$. Apply Problem 11g. □

## 11.5 Sub-Convex Recurrent Sequences

The investigations and results in the previous paragraphs cover the basic topic of the section, in the case of two or three successive terms involved in the recurrence relation. Does this mean that we have exhausted all the possibilities of this topic ? Not at all !

The following considerations are designated to show to the readers that no matter how deep and thorough our approach to a problem was, there still remain a lot of directions to be explored. The source problem will now be

**Problem 11h.** *Let $(a_n)_{n\geq 1}$ be a sequence of positive real numbers satisfying*

$$2a_{n+2} \leq a_{n+1} + a_n, \ \forall\, n \in \mathbb{N}^*.$$

*Prove that $(a_n)$ is convergent and compute its limit.*

**Solution.** By denoting $b_n = a_{n+1} + \frac{1}{2}a_n$, the inequality becomes

$$a_{n+2} + \frac{1}{2}a_{n+1} \leq a_{n+1} + \frac{1}{2}a_n \ , \ n \geq 1,$$

which shows that $(b_n)$ is a decreasing sequence (of positive real numbers). Hence $(b_n)$ is convergent. Let $b = \lim\limits_{n\to\infty} b_n$. Thus we have

$$\lim_{n\to\infty}\left(a_{n+1} + \frac{1}{2}a_n\right) = b$$

and by the result in Problem 11e (with $\lambda = -\frac{1}{2}$) we get

$$\lim_{n\to\infty} a_n = \frac{1}{1-\left(-\frac{1}{2}\right)} \cdot b = \frac{2b}{3},$$

which shows that $(a_n)$ is convergent and its limit is $\frac{2b}{3}$. □

We can solve in a similar manner the following more general problems.

**Problem 11i.** *If $(a_n)_{n\geq 1}$ is a sequence of positive real numbers and there exists $\lambda \in (0,1)$ such that*

$$a_{n+2} \leq (1-\lambda)a_{n+1} + \lambda a_n, \ \forall\, n \geq 1, \tag{8}$$

*then $(a_n)$ is convergent.*

**Problem 11j.** *If* $(x_n)$ *is a sequence of real positive numbers and there exist* $a, b \in (0,1)$, $a+b \leq 1$, *such that*

$$x_{n+2} \leq a x_{n+1} + b x_n, \ \forall\, n \geq 1, \tag{9}$$

*then* $(x_n)$ *is convergent.*

**Hints.** The inequality (8) can be written as

$$a_{n+2} + \lambda a_{n+1} \leq a_{n+1} + \lambda a_n, \ n \geq 1$$

and then we follow the same steps as in Problem 11h.

Problem 11j is now simply deduced, as $a + b \leq 1$ implies $b \leq 1 - a$; besides the sequence $(x_n)$ has positive real terms, so by (9) we get

$$x_{n+2} \leq a x_{n+1} + (1-a) x_n,$$

which is exactly the relation (8) with $\lambda = 1 - a$. □

## 11.6 The Case of $k$ Successive Terms

In this paragraph we will try to prove a more general result. Let $(x_n)$ be a sequence of real numbers and $(y_n)$ be the sequence defined by

$$y_n = b_0 x_n + b_1 x_{n+1} + \ldots + b_k x_{n+k}, \ n \geq 0, \tag{10}$$

where $b_0, b_1, \ldots, b_k$ are real constants such that

$$b_0 + b_1 + \ldots + b_k = 1. \tag{11}$$

We are searching for conditions to involve the coefficients $b_i$, $i = 1, 2, \ldots, k$, such that $\lim\limits_{n\to\infty} x_n = a \Leftrightarrow \lim\limits_{n\to\infty} y_n = a$.

It is clear that for $k = 1$ the answer is given by Problem 11e, as

$$\lim_{n\to\infty} (a_{n+1} - \lambda a_n) = (1-\lambda)a \Leftrightarrow \lim_{n\to\infty} \left( \frac{1}{1-\lambda} a_{n+1} - \frac{\lambda}{1-\lambda} a_n \right) = a,$$

(in this case we have $b_0 = -\dfrac{\lambda}{1-\lambda}$, $b_1 = \dfrac{1}{1-\lambda}$).

For $k = 2$, the answer is given by Problem 11g, since

$$\lim_{n\to\infty} (\lambda^2 a_{n+2} - 2\lambda a_{n+1} + a_n) = (\lambda - 1)^2 \cdot a$$

$$\Leftrightarrow \lim_{n\to\infty} \left( \frac{\lambda^2}{(\lambda-1)^2} \cdot a_{n+2} - \frac{2\lambda}{(\lambda-1)^2} \cdot a_{n+1} + \frac{1}{(\lambda-1)^2} \cdot a_n \right) = a$$

when the coefficients involved in the affine combination (10) are

$$b_0 = \frac{1}{(\lambda-1)^2}, \ b_1 = -\frac{2\lambda}{(\lambda-1)^2} \text{ and } b_2 = \frac{\lambda^2}{(\lambda-1)^2}.$$

It is rather difficult to guess the following property of the coefficients in the two previous particular cases: both equations $b_0t + b_1 = 0$ and $b_0t^2 + b_1t + b_2 = 0$ have a root (double, in the second case) $t_1$ satisfying $|t_1| < 1$.

The complete answer to our question is contained in

**Theorem 1.** *Let $k \in \mathbb{N}^*$ and $b_0, b_1, \ldots, b_k$ be real numbers satisfying (11). If $b_k \neq 0$ and the absolute values of all the roots of*

$$b_kt^k + b_{k-1}t^{k-1} + \ldots + b_1t + b_0 = 0 \tag{12}$$

*are less than 1, then*

$$\lim_{n\to\infty} x_n = a \Leftrightarrow \lim_{n\to\infty} (b_0x_n + b_1x_{n+1} + \ldots + b_kx_{n+k}) = a. \tag{*}$$

**Proof.** We proceed by induction with respect to $k$. For $k = 1$ and $k = 2$ the conclusion follows by Problems 11 and 11g, respectively.

We assume Theorem 1 holds for $k = m$ and prove that it also holds for $k = m + 1$. Let

$$b_0, b_1, \ldots, b_m, b_{m+1} \in \mathbb{R}, \quad b_{m+1} \neq 0,$$

such that

$$b_0 + b_1 + \ldots + b_m + b_{m+1} = 1.$$

We construct the sequence $(u_n)$ by

$$u_n = b_0x_n + b_1x_{n+1} + \ldots + b_{m+1}x_{n+m+1}, \quad n \geq 0$$

and let $\alpha_1, \alpha_2, \ldots, \alpha_{m+1}$ be the (possibly complex) roots of

$$b_{m+1}t^{m+1} + b_mt^m + \ldots + b_1t + b_0 = 0. \tag{12'}$$

We denote by $s_i$ $(i = 1, 2, \ldots, m)$ the symmetric elementary functions of variables $\alpha_1, \alpha_2, \ldots, \alpha_m$, that is

$$\begin{aligned} s_1 &= \alpha_1 + \alpha_2 + \ldots + \alpha_m, \\ s_2 &= \alpha_1\alpha_2 + \ldots + \alpha_1\alpha_m + \alpha_2\alpha_3 + \ldots + \alpha_{m-1}\alpha_m, \\ &\ \vdots \\ s_m &= \alpha_1\alpha_2\cdots\alpha_m, \end{aligned}$$

and by $S_j$ $(j = 1, 2, \ldots, m + 1)$ the symmetric elementary functions of variables $\alpha_1, \alpha_2, \ldots, \alpha_m, \alpha_{m+1}$. Let $(z_n)_{n\geq m}$ be defined by

$$z_n = x_{n+m} - s_1x_{n+m-1} + s_2x_{n+m-2} + \ldots + (-1)^m s_mx_n.$$

We have

$$\begin{aligned} z_{n+1} - \alpha_{m+1}z_n &= x_{n+m+1} - s_1x_{n+m} + s_2x_{n+m-1} + \ldots + (-1)^m s_mx_{n+1} \\ &\quad -\alpha_{m+1}\left[x_{n+m} - s_1x_{n+m-1} + s_2x_{n+m-2} + \ldots + (-1)^m s_mx_n\right] \\ &= x_{n+m+1} - (s_1 + \alpha_{m+1})\,x_{n+m} + (s_2 + \alpha_{m+1}s_1)\,x_{n+m-1} + \ldots \\ &\quad +(-1)^m(s_m + \alpha_{m+1}s_{m-1})x_{n+1} + (-1)^{m+1}\alpha_{m+1}s_mx_n \\ &= x_{n+m+1} - S_1x_{n+m} + S_2x_{n+m-1} + \ldots + (-1)^{m+1}S_{m+1}x_n. \end{aligned}$$

As $\alpha_1, \alpha_2, \ldots, \alpha_m, \alpha_{m+1}$ are the roots of equation (12′), we have

$$S_j = (-1)^j \cdot \frac{b_{m+1-j}}{b_{m+1}}, \quad j = 1, 2, \ldots, m+1,$$

and hence

$$z_{n+1} - \alpha_{m+1} z_n = x_{m+n+1} + \frac{b_m}{b_{m+1}} \cdot x_{n+m} + \frac{b_{m-1}}{b_{m+1}} x_{n+m-1} + \ldots + \frac{b_0}{b_{m+1}} \cdot x_n.$$

This can be written using the expression of $(u_n)$ as

$$z_{n+1} - \alpha_{m+1} z_n = \frac{1}{b_{m+1}} \cdot u_n. \tag{13}$$

By hypothesis we know that $\lim_{n\to\infty} u_n = a$. Hence by (13) it results that

$$\lim(z_{n+1} - \alpha_{m+1} z_n) = \frac{1}{b_{m+1}} \cdot a.$$

Since $\alpha_{m+1}$ is a root of equation (12′) and hence it fulfills the assumptions in the hypothesis, it results that

$$|\alpha_{m+1}| < 1.$$

Using now Problem 11e, we get

$$\lim_{n\to\infty} z_n = \frac{1}{1-\alpha_{m+1}} \cdot \frac{a}{b_{m+1}},$$

which is equivalent to

$$\lim_{n\to\infty} (1-\alpha_{m+1}) b_{m+1} z_n = a. \tag{14}$$

The obtained sequence in the left-hand side of (14) is defined by means of a linear combination of $n+1$ successive terms of $(x_n)$. Moreover, the equation associated to that sequence is

$$b_{m+1}(1-\alpha_{m+1}) \left[t^m - s_1 t^{m-1} + s_2 t^{m-2} + \ldots + (-1)^m s_m\right] = 0 \tag{15}$$

which is equivalent to

$$t^m - s_1 t^{m-1} + \ldots + (-1)^m s_m = 0, \tag{15′}$$

due to the fact that $b_{m+1} \neq 1$ and $|\alpha_{m+1}| < 1$.

But the roots of (15′), $\alpha_1, \alpha_2, \ldots, \alpha_m$, are by hypothesis less than 1 in absolute value. Moreover, the coefficients of the linear combination in (14) have the sum 1, as

$$\begin{aligned} &b_{m+1}(1-\alpha_{m+1})\left[1 - s_1 + s_2 - \ldots + (-1)^m s_m\right] \\ &= b_{m+1}\left[1 - S_1 + S_2 - \ldots + (-1)^m S_m\right] \\ &= b_{m+1} \cdot \frac{b_0 + b_1 + \ldots + b_{m+1}}{b_{m+1}} = 1. \end{aligned}$$

Therefore, the sequence $((1-\alpha_{m+1})b_{m+1}z_n)_{n\geq 0}$ satisfies all the assumptions in Theorem 1. Using the induction assumption (i.e., the fact that (*) holds for $k=m$), from

$$\lim_{n\to\infty}(1-\alpha_{m+1})b_{m+1}\left[x_{n+m}-s_1x_{n+m-1}+\ldots+(-1)^m s_m x_n\right]=a$$

it results that

$$\lim_{n\to\infty} x_n = a.$$

We actually proved that

$$\lim_{n\to\infty}(b_0x_n+b_1x_{n+1}+\ldots+b_{m+1}x_{n+m+1})=a \Leftrightarrow \lim_{n\to\infty}x_n=a,$$

that is, the conclusion holds for $k=m+1$ as well. □

**Remark.** If the absolute value of the roots of (12) are not strictly less than 1, the conclusion of Theorem 1 is not valid anymore, as shown by the next example.

**Example.** Let $(x_n)$ be the sequence defined by

$$x_{4n}=0,\ x_{4n+1}=-1,\ x_{4n+2}=2,\ x_{4n+3}=3,\ n\geq 0.$$

If we take $k=2, b_0=\dfrac{1}{2}, b_1=0$ and $b_1=\dfrac{1}{2}$ and construct the sequence $(u_n)_{n\geq 0}$ by

$$u_n=b_0x_n+b_1x_{n+1}+b_2x_{n+2},\ n\geq 0,$$

then we have

$$u_n=1,\ \ \forall\, n\geq 0$$

and hence

$$\lim_{n\to\infty}(b_0x_n+b_1x_{n+1}+b_2x_{n+2})=1,$$

but $\lim\limits_{n\to\infty}x_n$ does not exist. Here the associated equation, $\dfrac{1}{2}t^2+\dfrac{1}{2}=0$, has the roots of absolute value 1 (and not less than 1).

Two remarkable particular cases of Theorem 1 are given by

**Problem 11k.** *The quadratic equation $ax^2+bx+c=0$ has its roots in the interval $(-1,1)$ and $(x_n)_{n\geq 1}$ is a real sequence such that*

$$\lim_{n\to\infty}(ax_{n+1}+bx_n+cx_{n-1})=0.$$

*Prove that*

$$\lim_{n\to\infty}x_n=0.$$

**Problem 11l.** *Let $(y_n)_{n\geq 1}$ be a real sequence such that*

$$\lim_{n\to\infty}(8y_{n+1}+4y_n-y_{n-1})=11.$$

*Prove that*

$$\lim_{n\to\infty}y_n=1.$$

## 11.7 Bibliographical Comments

The source problem in this section was proposed at the Mathematical Olympiad in U.S.S.R. in 1977. It was also published in [1], p. 42, as problem 1.15, where a solution of it is given. The source problem has been extended by Gh. Miculescu in *Gazeta Matematica* no. 7/1980, problem 18351*, which is exactly Problem 11a in our book.

Problem 11b can be found in [2], pp. 76–77, Application 9.1.2, as well as Problem 11c, which is there Problem 2.6.45, p. 24.

Problem 11f was proposed by the author first at the Intercounty Mathematics Competition "Gr.C. Moisil", $2^{nd}$ edition, Bistrita, 1987, and then published in *Revista Matematica a Elevilor din Timisoara*, no. 1/1987, as problem 6140.

Problem 11g is a reply to Problem 11f and was published in the same journal, as problem 6141, author Dorel Mihet.

Problem 11h is a particular case of a class of recurrent sequences that appear in relation with proving several generalized fixed point theorems, see [9], for example. The most important results of this kind are due to Presic (1965), Copson (1970), Russel (1972) and so on. For the Romanian mathematical literature, we refer to [10]–[13], where various particular cases are considered.

The content of paragraph 11.6 is taken from our paper [4]. Problems 11k and 11l are, respectively, part i) and ii) of problem C:2136 proposed by C. Mortici in *Gazeta Matematica* no. 2/1999.

## 11.8 References


1. **D.M. Batinetu,** *Sequences.* (Romanian), Editura Albatros, Bucuresti, 1979

2. **D.M. Batinetu et al.** *Exercises and Problems of Analysis for XI and XII Forms.* (Romanian), Editura Didactica si Pedagogica, Bucuresti, 1981

3. **D.M. Batinetu et al.,** *Analysis. Exercises and Problems.* (Romanian), Editura Militara, Bucuresti, 1992

4. **V. Berinde,** *Characterizations of Convergent Sequences by Means of Linear Combinations Containing its k Successive Terms* (Romanian), Lucr. Semin. Creativ. Mat., **1** (1991–1992), 25–38

5. **C. Forga,** *The Computation of the Limits of Some Sequences Defined by Recurrence Relations.* (Romanian), Revista Matematica a Elevilor din Timisoara, **20** (1989), no. 1–2, 39–42

6. **M. Megan, P. Preda,** *Recurrent Real Sequences.* (Romanian), Caiete metodico-stiintifice, Univ. Timisoara, no. 20, 1984

7. **D. Mihet, M. Piticari,** *A Convergence Problem.* (Romanian), Gazeta Matematica (Serie metodica), Anul **11 (90)** (1990), no. 1, 30–31

8. **I. Prodanov et al.**, *Zbornik ot Zadaci po Diferentialnov i Integralno Smiatone Izd. Nauka i iskustow, Sofia, 1976*

9. **M. Taskovic,** *Osnove teorije fiksne tacke.* Zavod za Udzbenike i Nastavna Sredstva, Beograd, 1986

10. **D. Barbosu,** *On Some Sub-convex Recurrences.* (Romanian), Lucr. Semin. Creativ. Mat., **3** (1993–1994), 53–60

11. **D. Barbosu,** *Systems of Sub-convex Sequences of First Order.* (Romanian), Lucr. Semin. Creativ. Mat., **5** (1995-1996), 25–28

12. **D. Barbosu, M. Andronache,** *On the Convergence of Sub-convex Sequences.* (Romanian), Gazeta Matematica, **102** (1997), no. 1, 3–4

13. **A. Sandovici,** *Two Classes of Almost Convex Sequences.* (Romanian), Lucr. Semin. Creativ. Mat., **9** (1999-2000), 157–168

# Theme 12

# Binomial Characterizations of Arithmetic Progressions

By definition, a sequence of numbers $(a_p)_{p\geq 1}$ is an arithmetic progression if, beginning with the second one, each term is the arithmetic mean of its two closest neighbors, i.e.,

$$a_{k+1} = \frac{a_k + a_{k+2}}{2}, \quad k \geq 1. \tag{1}$$

The difference of two consecutive terms of the progression is a constant which is called the "ratio" of the arithmetic progression and is usually denoted by $r$ :

$$a_2 - a_1 = a_3 - a_2 = \ldots = r. \tag{2}$$

Each of the relations (1) and (2) characterizes the arithmetic progression $(a_p)_{p\geq 1}$, which is also denoted by $(a_p) \div$ . We can write (1) equivalently as

$$a_k - 2a_{k+1} + a_{k+2} = 0, \quad k \geq 1, \tag{1'}$$

which is a linear combination of three consecutive terms of the given arithmetic progression. If four consecutive terms of the progression are involved, then we obtain the next problem, which is exactly the source problem of the present chapter.

## 12.1 Source Problem

**Problem 12.** *If* $(a_p)_{p\geq 1}$ *is an arithmetic progression, then*

$$a_p - 3a_{p+1} + 3a_{p+2} - a_{p+3} = 0 \tag{3}$$

*holds for each* $p \geq 1$.

**Solution.** By taking $k = p$ and $k = p + 1$ in (1), we have that

$$a_p - 2a_{p+1} + a_{p+2} = 0, \text{ and } a_{p+1} - 2a_{p+2} + a_{p+3} = 0,$$

respectively. Now subtracting these two relations side by side we obtain exactly (3).

It is easy to notice that in (1′) and (3) the binomial coefficients (1,2,1) and (1,3,3,1), respectively, appear with alternating signs. A question arises naturally: do such relations hold for any number of consecutive terms of the progression? The answer is given by

**Problem 12a.** *If* $(a_n)_{n\geq 1}$ *is an arithmetic progression, show that*

$$a_p - \binom{n}{1}a_{p+1} + \binom{n}{2}a_{p+2} - \ldots + (-1)^n\binom{n}{n}a_{p+n} = 0 \tag{4}$$

*holds for any* $n, p \in \mathbb{N}^*$, $n \geq 2$.

**Solution.** Let $r$ be the ratio of the progression. As

$$a_{p+i} = a_p + i \cdot r,$$

for each $i \in \mathbb{N}$, (4) can be written as

$$a_p\left[1 - \binom{n}{1} + \binom{n}{2} - \ldots + (-1)^n\binom{n}{n}\right]$$
$$-r\left[\binom{n}{1} - 2\binom{n}{2} + \ldots + (-1)^n n\binom{n}{n}\right] = 0.$$

It is easy to notice that

$$1 - \binom{n}{1} + \binom{n}{2} - \ldots + (-1)^n\binom{n}{n} = (1-1)^n = 0 \tag{5}$$

and we still have to prove that

$$\binom{n}{1} - 2\binom{n}{2} + \ldots + (-1)^n n\binom{n}{n} = 0. \tag{5'}$$

Now using the recurrence formula for combinations

$$k\binom{n}{k} = n\binom{n-1}{k-1}, \quad n \geq 2, \ k \geq 1,$$

(5′) will be easily written as an obvious relation

$$n\left[\binom{n-1}{0} - \binom{n-1}{1} + \ldots + (-1)^{n-1}\binom{n-1}{n-1}\right] = 0.$$

**Remark.** Once we have solved Problem 12, another question occurs: is relation (4) characteristic to the arithmetic progressions? That is, does the converse of the statement in Problem 12a hold? The answer is given by

**Problem 12b.** *Show that the terms of the sequence* $(a_n)_{n\geq 1}$ *defined by*

$$a_{n+1} = (-1)^{n+1}\left[a_1 - \binom{n}{1}a_2 + \binom{n}{2}a_3 - \ldots + (-1)^{n-1}\binom{n}{n-1}a_n\right] \quad (6)$$

*form an arithmetic progression.*

**Solution.** It is easy to notice that (6) can be written as

$$a_1 - \binom{n}{1}a_2 + \binom{n}{2}a_3 - \ldots + (-1)^n\binom{n}{n}a_{n+1} = 0 \quad (6')$$

which is exactly (4) with $p = 1$. We are going to use the induction method now to prove the given statement.

For $n = 2$, from $(6')$ we have that $a_1 - 2a_2 + a_3 = 0$, i.e., $a_1, a_2, a_3 \div$ . Now we suppose that $a_1, a_2, \ldots, a_k \div$ is true and try to show that we have also $a_1, a_2, \ldots, a_k, a_{k+1} \div$ .

From $(6')$ we obtain, for $n = k$,

$$a_1 - \binom{k}{1}(a_1 + r) + \binom{k}{2}(a_1 + 2r) - \ldots + (-1)^{k-1}\binom{k}{k-1}\cdot[a_1 + (k-1)r]$$

$$+(-1)^k\binom{k}{k}a_{k+1} = 0,$$

where $r$ is the ratio of the arithmetic progression $a_1, a_2, \ldots, a_k$. This last relation is equivalent to

$$a_1\left[1 - \binom{k}{1} + \binom{k}{2} - \ldots + (-1)^{k-1}\binom{k}{k-1}\right]$$

$$-r\left[\binom{k}{1} - 2\binom{k}{2} + \ldots + (-1)^k(k-1)\binom{k}{k-1}\right] + (-1)^k C_k^k a_{k+1} = 0. \quad (7)$$

We know by (5) that

$$1 - \binom{k}{1} + \binom{k}{2} - \ldots + (-1)^k\binom{k}{k-1} = (-1)^{k+1}\binom{k}{k}$$

and by $(5')$ that

$$\binom{k}{1} - 2\binom{k}{2} + 3\binom{k}{3} - \ldots + (-1)^k(k-1)\binom{k}{k-1} = (-1)^k\cdot k\binom{k}{k}.$$

By replacing these in (7) we obtain

$$a_1\cdot(-1)^{k+1} + r\cdot(-1)^{k+1}\cdot k + (-1)^k\cdot a_{k+1} = 0$$

that leads us to

$$a_{k+1} = a_1 + k \cdot r,$$

which shows that $a_1, a_2, \ldots, a_k, a_{k+1} \div$ .

Now, according to the principle of mathematical induction, $(a_n)_{n\geq 1}$ is an arithmetic progression. □

**Remark.** The relations (5) and (5′) make us think about their duals

$$1 + \binom{n}{1} + \binom{n}{2} + \ldots + \binom{n}{n} = 2^n, \tag{5''}$$

respectively

$$\binom{n}{1} + 2\binom{n}{2} + \ldots + n\binom{n}{n} = n \cdot 2^{n-1}. \tag{5'''}$$

So, if $(a_p)_{p\geq 1}$ is an arithmetic progression with the ratio $r$, using (5′) and (5″) and following the same steps as in the solution of Problem 11a, we obtain that

$$\begin{aligned}
&a_p + \binom{n}{1}a_{p+1} + \binom{n}{2}a_{p+2} + \ldots + \binom{n}{n}a_{p+n} \\
&= a_p\left[1 + \binom{n}{1} + \ldots + \binom{n}{n}\right] + +r\left[\binom{n}{1} + 2\binom{n}{2} + \ldots + n\binom{n}{n}\right] \\
&= a_p \cdot 2^n + +r \cdot n \cdot 2^{n-1} = 2^{n-1}(2a_p + r \cdot n) = 2^{n-1}\left[a_p + (a_p + n \cdot r)\right] \\
&= 2^{n-1} \cdot (a_p + a_{p+n}) = 2^n \cdot \frac{a_p + a_{p+n}}{2}.
\end{aligned}$$

This result can be formulated as follows.

**Problem 12c.** *If $(a_p)_{p\geq 1}$ is an arithmetic progression, then*

$$\frac{\binom{n}{0}a_p + \binom{n}{1}a_{p+1} + \ldots + \binom{n}{n}a_{p+n}}{\binom{n}{0} + \binom{n}{1} + \ldots + \binom{n}{n}} = \frac{a_p + a_{p+n}}{2}. \tag{8}$$

**Remark.** There is a very nice interpretation of Problem 12c: the weighted mean of $n+1$ consecutive terms of an arithmetic progression having

$$\binom{n}{0}, \binom{n}{1}, \ldots, \binom{n}{n}$$

as weights is equal to the arithmetic mean of the extreme terms involved.

A question arises now: could this be true, that any sequence of numbers having the property above, for any $n, p \in \mathbb{N}$, is an arithmetic progression? The answer lies in

**Problem 12d.** *If the terms of the sequence $a_p, a_{p+1}, \ldots$ satisfy (8) for any $n \in \mathbb{N}^*$, $n \geq 2$, then $(a_p)_{n\geq p}$ is an arithmetic progression.*

**Solution.** First we consider the case $p = 1$. Then (8) will be written as

$$a_1 + \binom{n}{1} a_2 + \ldots + \binom{n}{n} a_{n+1} = 2^{n-1} \cdot (a_1 + a_{n+1}). \tag{8'}$$

We are going to show now, by induction, that $a_1, a_2, \ldots, a_n, a_{n+1} \div$. For $n = 2$, from (8′) we obtain that

$$a_1 + 2a_2 + a_3 = 2(a_1 + a_3) \Leftrightarrow 2a_2 = a_1 + a_3$$

and so $a_1, a_2, a_3 \div$.

Now we assume $a_1, a_2, \ldots, a_n \div$ and try to show that $a_1, a_2, \ldots, a_n, a_{n+1} \div$.

If $r$ is the ratio of the progression $a_1, a_2, \ldots, a_n$, then from (8′) we obtain that

$$a_1 + \binom{n}{1}(a_1 + r) + \binom{n}{2}(a_1 + 2r) + \ldots + \binom{n}{n-1} \cdot (a_1 + (n-1)r) + \binom{n}{n} a_{n+1}$$
$$= 2^{n-1} \cdot (a_1 + a_{n+1})$$
$$\Leftrightarrow a_1 \left[1 + \binom{n}{1} + \ldots + \binom{n}{n-1}\right] + r \cdot \left[\binom{n}{1} + 2\binom{n}{2} + \ldots + (n-1)\binom{n}{n-1}\right]$$
$$+ a_{n+1} = 2^{n-1}(a_1 + a_{n+1}).$$

Using the relations (5″) and (5‴) we obtain

$$a_1(2^n - 1) + r\,(n \cdot 2^{n-1} - n) - a_1 \cdot 2^{n-1} = a_{n+1}(2^{n-1} - 1)$$

that leads us easily to

$$a_{n+1} = a_1 + n \cdot r,$$

and so $a_1, a_2, \ldots, a_n, a_{n+1} \div$. □

Coming back to Problem 12a, this can suggest us a stronger result:

**Problem 12e.** *If $(a_n)_{n \geq 1}$ is an arithmetic progression, then*

$$a_p^k - \binom{n}{1} a_{p+1}^k + \binom{n}{2} a_{p+2}^k - \ldots + (-1)^n \binom{n}{n} a_{p+n}^k = 0 \tag{9}$$

*holds for any $n, p, k \in \mathbb{N}^*, n \geq 3$.*

**Solution.** We are going to use the method of induction on $k$. For $k = 1$ the relation (9) is the same as (4), so it holds for any $n, p \in \mathbb{N}^*, n \geq 2$. Now we assume that (9) holds for a certain $k$ and we prove that it also holds for $k + 1$, i.e.,

$$a_p^{k+1} - \binom{n}{1} a_{p+1}^{k+1} + \binom{n}{2} a_{p+2}^{k+1} - \ldots + (-1)^n \binom{n}{n} a_{p+n}^{k+1} = 0,$$
$$\forall\, n, p \in \mathbb{N}^*.$$

Indeed, this last relation is equivalent to

$$a_p^k \cdot a_p - \binom{n}{1} a_{p+1}^k \cdot a_{p+1} + \ldots + (-1)^n \binom{n}{n} a_{p+n}^k \cdot a_{p+n} = 0$$

and, taking into account that $a_p, a_{p+1}, \ldots, a_{p+n}$ is an arithmetic progression (let $r$ be its ratio), we have that

$$a_p \left[ a_p^k - \binom{n}{1} a_{p+1}^k + \binom{n}{2} a_{p+2}^k - \ldots + (-1)^{n+1} \binom{n}{n} a_{p+n}^k \right]$$

$$-r \left[ \binom{n}{1} a_{p+1}^k - 2\binom{n}{2} a_{p+2}^k + \ldots + (-1)^{n+1} n \binom{n}{n} a_{p+n}^k \right] = 0. \tag{10}$$

As the first square bracket in (10) is equal to zero, according to the induction hypothesis, we still have to prove that

$$\binom{n}{1} a_{p+1}^k - 2\binom{n}{2} a_{p+2}^k + \ldots + (-1)^{n+1} n \binom{n}{n} a_{p+n}^k = 0,$$

which, taking into account that $k\binom{n}{k} = n \cdot \binom{n-1}{k-1}$, is equivalent to

$$n \left[ a_{p+1}^k - \binom{n-1}{1} a_{p+2}^k + \ldots + (-1)^{n-1} \binom{n-1}{n-1} a_{p+n}^k \right] = 0.$$

The last relation is exactly (9), with $p+1$ instead of $p$ and $n-1$ instead of $n$. According to the induction hypothesis again, this holds for any $n - 1 \geq 3$, i.e., for any $n \geq 4$. For $n = 3$ the relation (1) can be proved directly. So (9) holds for any $n, p, k \in \mathbb{N}^*, n \geq 3$. □

Now combining Problems 12a and 12e we are led to

**Problem 12f.** *If $(a_p)_{p\geq 1}$ is a sequence for which the relation (4) holds, for any $n \in \mathbb{N}^*, n \geq 2$, then the relation (9) holds also for any $k, n \in \mathbb{N}^*, n \geq 3$.*

**Remark.** For $p = 1$ (and $x$ instead of $a$), from Problem 12f we obtain

**Problem 12g.** *If $(x_n)_{n\geq 1}$ is a sequence of numbers for which*

$$x_1 - \binom{n}{1} x_2 + \binom{n}{2} x_3 - \ldots + (-1)^n \binom{n}{n} x_{n+1} = 0$$

*holds for any $n \in \mathbb{N}, n \geq 2$, then for each $k \in \mathbb{N}^*$*

$$x_1^k - \binom{n}{1} x_n^k + \binom{n}{2} x_3^k - \ldots + (-1)^n \binom{n}{n} x_{n+1}^k = 0$$

*holds also for any $n \in \mathbb{N}, n \geq k+1$.*

**Remarks.** $1^0$. The condition $n \geq k+1$ in Problem 12g is crucial. For example, if we take $n = 3, k = 3$, then the equality $x_1^3 - 3x_2^3 + 3x_3^3 - x_4^3 = 0$ does not hold anymore, although $x_1, x_2, x_3, x_4 \div$ .

$2^0$. By analyzing Problems 12f and 12g we naturally come to the following problem:

*Find all the sets $K \subset \mathbb{N}^*$ which have the property that if (9) holds for $k \in K$, then (9) holds for any $n \in \mathbb{N}^*, n \geq 3$.*

Taking into account the previous results, it is obvious that $K = \{1\}$ is a solution of the problem above. But are there any other solutions?

In the same context we invite readers to search on the following open problem.

**Problem 12h.** *Assuming that $(a_n)_{n\geq 1}$ is a sequence of numbers, does the relation*

$$a_p^2 - \binom{n}{1} a_{p+1}^2 + \binom{n}{2} a_{p+2}^2 + \ldots + (-1)^n \binom{n}{n} a_{p+n}^2 = 0$$

*being valid for any $n, p \in \mathbb{N}^*, n \geq 3$ imply or not that $(a_n) \div$ ?*

At the end of this section we present another binomial characterization of the arithmetic progression.

**Problem 12i.** *Let $a_1, a_2, \ldots, a_n$ be an arithmetic progression with the ratio $r$ and*

$$S_{m,n} = a_1^m + a_2^m + \ldots + a_n^m.$$

*a) Show that*

$$a_{n+1}^p - a_1^p = \binom{p}{1} r\, S_{p-1,n} + \binom{p}{2} r^2\, S_{p-2,n} + \ldots + \binom{p}{p} r^p\, S_{0,n},$$

*for any naturals $p$ and $n$;*
*b) Conversely, show that if the identity at a) holds for any naturals $p$ and $n$, then the sequence $(a_n)_{n\in\mathbb{N}}$ is an arithmetic progression with the ratio $r$.*

We leave the solution of this problem to the readers, inviting them to try to formulate some other problems on this theme, too, in the same way we did starting from the source problem.

## 12.2 Bibliographical Comments

Problem 12a was published in *Gazeta Matematica*, no. 7/1991 as problem C:1153, having V. Berinde as author. Problem 12b is a reciprocal of Problem 12a and was published in *Gazeta Matematica* no. 11–12/1991 as problem C:1207, having I. Danila as author. From Problem 12e we obtain, for $p = 1$, Problem 16.3, p. 459 in [2]. Problem 12g is the $3^{rd}$ problem proposed by the author at the National Mathematics Olympiad, Arad, 1994, for the $10^{th}$ grade.

Problem 12i is contained in [3], p. 38, as problem 1.8.45 (with a shorter statement).

## 12.3 References

1. **C. Nastasescu et al.,** *Algebra. Textbook for 10$^{th}$ form.* (Romanian), Editura Didactica si Pedagogica, 1992

2. **M. Becheanu et al.,** *Problems of Algebra, Calculus and Geometry* (Romanian), Editura Cartea Romaneasca, 1991

3. **A.V. Leonte, C.P. Niculescu,** *Problems Book in Algebra and Calculus.* (Romanian), Editura Scrisul Romanesc, 1992

# Theme 13

# Using Duality in Studying Homographic Recurrences

## 13.1 Introduction

Sequences, and in a more particular way, recurrent sequences, offer a favorite medium for the first creative attempts of high school students and young mathematics teachers. Even though the theoretical bases for this domain have been established long ago and are well represented in mathematical publications, for students — whose mathematical knowledge is now being formed — problems in this subject can cause many surprises and difficulties, as well as they can be fertile ground for their first investigations.

The homographic recurrences hold a distinct place among this type of problems, because of the numerous geometrical properties of the iteration function by which they are defined (homographic function), properties that exhibit attractive and inciting qualities, stimulating the reader in the effort of (re)discovering and (re)valuing them by himself. A good reference for this is Chapter 1 of the book [1], pp. 1–68. Here we find the necessary theoretical aspects of the homographic function and detailed applications of it in the study of recurrent homographic sequences, applications that cover a fairly wide area of this theme.

Another method of approaching homographic recurrences is presented in the general case and then applied for two alternative examples in the book [5].

Synthesizing these two works, we could be tempted to say that the veins of this subject have been exhausted long ago, but we would be wrong.

The aim of this section, besides the one of the stated theme, is to show — once again — that in mathematics we can rarely declare the creative resources of a research domain as being exhausted. We want to show the reader that even where things seem to be carefully sorted and well arranged, there still is room for new explorations and investigations, after which we get a little but well-deserved award: new results. We hope that with these new results, as well as with the open problems at the end of the theme, we will stimulate the creativity of youngsters, high school or university students.

We have to mention that in the present section we will not refer to the general properties of the homographic function, but we will present a simple and much more accessible method of approaching the homographic recurrent sequences.

## 13.2 Source Problem

The starting point of all investigations in this section is

**Problem 13.** *Prove that any sequence $(x_n)_{n\geq 0}$ of real numbers with the property $x_0 > 0$ and $x_{n+1}x_n - 2x_n = 3$, $(\forall)\, n \in \mathbb{N}$, is convergent.*

**Solution.** We will now present three methods, taken from [2], where we solved a more general problem.

**Solution 1.** Because $x_0 > 0$ and the relation in the statement can be written as

$$x_{n+1}x_n = 2x_n + 3,$$

we can easily prove by induction that

$$x_n > 2, \quad n \in \mathbb{N}. \tag{1}$$

This ensures that the relation in the problem can be written as

$$x_{n+1} - 3 = \frac{-1}{x_n}(x_n - 3),\ n \geq 0,$$

so we can obtain

$$|x_{n+1} - 3| = \frac{1}{x_n}\,|x_n - 3|\,,\ n \geq 0. \tag{2}$$

Using the inequality (1) and the relation (2) we get

$$|x_{n+1} - 3| < \frac{1}{2}\,|x_n - 3|\,,\ n \geq 0,$$

and, by induction,

$$|x_n - 3| < \left(\frac{1}{2}\right)^n \cdot |x_0 - 3|\,,\ n \geq 1,$$

which shows, using the majorizing criterion, that $\lim\limits_{n\to\infty} x_n = 3$. □

**Solution 2.** Considering inequality (1), the relation in the statement can be written as

$$x_n = \frac{3}{x_{n+1} - 2},\ n \geq 0.$$

The same relation (1) allows us to use the substitution

$$y_n = \frac{1}{x_n + 1}, \; n \geq 0,$$

and hence

$$y_n = \frac{1}{x_n + 1} = \frac{1}{\dfrac{3}{x_{n+1} - 2} + 1} = \frac{x_{n+1} - 2}{x_{n+1} + 1} = 1 - 3y_{n+1},$$

which shows that the sequence $(y_n)$ satisfies the linear nonhomogeneous recurrence relation

$$y_{n+1} = -\frac{1}{3}y_n + \frac{1}{3}, \quad n \geq 0.$$

Solving this recurrence (e.g., by dividing each term by $\left(-\frac{1}{3}\right)^{n+1}$ and then adding up the relations obtained for $n := 0, 1, 2, \ldots, n-1$) we get

$$y = \left(-\frac{1}{3}\right)^{n-1} \cdot y_1 + \frac{1}{4}\left(1 - \left(-\frac{1}{3}\right)^{n-1}\right), \quad n \geq 1.$$

Turning the previous relation to a limit we get

$$\lim_{n\to\infty} y_n = \frac{1}{4}$$

and so

$$\lim_{n\to\infty} x_n = \frac{1}{\lim\limits_{n\to\infty} y_n} - 1 = 3,$$

which finishes the solution. □

**Solution 3.** We suppose that $x_0 < 3$ (when $x_0 > 3$ we tackle it similarly and if $x_0 = 3$ we show by induction that $x_n = 3$, $\forall n \geq 0$, and so the sequence is convergent). We can easily prove by induction that

$$x_0 \leq x_n \leq x_1, \quad n \geq 0,$$

and therefore the sequence $(x_n)$ is bounded.

We can also prove the inequalities

$$x_{2n} < 3 \text{ and } x_{2n+1} > 3, \quad n \geq 0. \tag{3}$$

The relation in the statement leads to

$$x_{n+2} - x_n = -\frac{2(x_n - 3)(x_n + 1)}{2x_n + 3},$$

which, together with (1) and (3), shows that the subsequence $(x_{2n})_{n\geq 0}$ is strictly increasing, and the subsequence $(x_{2n+1})_{n\geq 0}$ is strictly decreasing. Therefore, both subsequences are convergent. Let

$$u = \lim_{n\to\infty} x_{2n} \text{ and } v = \lim_{n\to\infty} x_{2n+1}$$

and take in the recurrence relation $n$ even and $n$ odd, respectively. Passing to the limit each time, we get

$$uv - 2u = 3 \text{ and } uv - 2v = 3$$

respectively, from which we obtain firstly $u = v$ and then $u = 3$. Thus, $(x_n)$ is convergent and its limit is 3. □

## 13.3 A Close Examination of the Solution. First Investigations

Problem 13 was solved in the previous section under the essential assumption that $x_0 > 0$.

Writing the recurrence relation as

$$x_{n+1}x_n - [(3-1)]\, x_n + 3\cdot(-1) = 0,$$

this suggests we might consider the more general case in which 3 and $-1$ are replaced by two arbitrary numbers $a$ and $b$. We therefore get

**Problem 13a.** *Show that if $a, b \in \mathbb{R}_+^*$ and $2b < a$, then the sequence $x_{n+1}\cdot x_n + (b-a)\,x_n = ab$, $n \geq 0$, is convergent for any $x_0 > 0$.*

**Solution.** We follow exactly the same steps as in solving Problem 13, in order to show that the limit is $a$, see [2]. □

Coming back to Problem 13, let us compare the three methods. Each of them has its own beauty, its limits and a certain degree of generality which will allow us to apply the three of them in solving Problem 13a, a much more general one.

First, we will be concerned with the first method. The essential point in applying this method was that the sequence $(x_n)$ has 2 (a number greater than 1) as the lower bound. Would we obtain the same kind of sequence if $x_0 > 0$ were weakened ?

Because $x_{n+1} = \dfrac{2x_n+3}{x_n}$, it is easy to notice that if $x_0 < -\dfrac{3}{2}$, then $\dfrac{2x_0+3}{x_0} > 0$, i.e., $x_1 > 0$, which ensures that the sequence has a lower bound

$$x_n > 2, \forall n \geq 2. \qquad (1')$$

Therefore, from (2) we will now get

$$|x_{n+1} - 3| < \frac{1}{2}\,|x_n - 3|\,,\ n \geq 2$$

and so the convergence of the sequence will result from the inequality

$$|x_{n+1} - 3| < \left(\frac{1}{2}\right)^{n-2} \cdot |x_2 - 3|, n \geq 2.$$

In this way we just solved

**Problem 13b.** *The sequence* $(x_n)$ *defined by the recurrence relation*

$$x_{n+1} \cdot x_n - 2x_n = 3, \quad n \geq 0 \tag{4}$$

*is convergent, for any* $x_0 \in \left(-\infty, -\frac{3}{2}\right) \cup (0, \infty)$.

In fact, even this problem can be improved to get

**Problem 13c.** *The sequence* $(x_n)$ *defined by* (4) *is convergent for any*

$$x_0 \in \left(-\infty, -\frac{3}{2}\right) \cup \left(-\frac{6}{7}, 0\right) \cup (0, \infty).$$

**Solution.** It can be verified immediately that for $n \geq 2$ we get $x_n > 0$ and thus relation (4) will hold for any $n \geq 3$. The convergence will now result from the inequality

$$|x_n - 3| < \left(\frac{1}{2}\right)^{n-3} \cdot |x_3 - 3|,$$

that holds for any $n \geq 4$. □

**Remark.** At this point we must highlight the successive extensions of the domain of values for the first term of the sequence: from $(0, \infty)$ to $\left(-\infty, -\frac{3}{2}\right) \cup (0, \infty)$ and then to $\left(-\infty, -\frac{3}{2}\right) \cup \left(-\frac{6}{7}, 0\right) \cup (0, \infty)$. This suggests the following question: are the restrictions imposed for $x_0$ really necessary? Which means, couldn't we have, for example, $x_0 = -\frac{3}{2}$ or $x_0 = -\frac{6}{7}$ ? The answer is immediate: if we start off with $x_0 = -\frac{6}{7}$, then writing the recurrence relation as

$$x_{n+1} = 2 + \frac{3}{x_n} \tag{4'}$$

we obtain $x_1 = -\frac{3}{2}$, $x_2 = 0$ and so $x_3$ wouldn't be defined, because the fraction in (4′) would have the denominator zero.

The example above proves that restricting the values for $x_0$ we tend to ensure a correct definition for the given recurrence.

Now we get to the next question: does the whole interval $\left[-\frac{3}{2}, -\frac{6}{7}\right]$ consist of not admitted values for $x_0$? If not, how can we describe these values?

We will try to answer both these questions in the next paragraph.

## 13.4 Dual Homographical Sequences

In order to simplify the notation, we use the fact that any homographic recurrence

$$x_{n+1} = \frac{ax_n + b}{cx_n + d}, \quad a, b, c, d \in \mathbb{R}, \ ad - bc \neq 0$$

can be reduced to a recurrence of the form

$$u_{n+1}u_n + \alpha u_n + \beta = 0, \tag{5}$$

by substituting $u_n = cx_n + d$.

Because of this, we will deal from now on with the reduced form (5) of the homographic recurrences.

**Definition 1.** The equation $r^2 + \alpha r + \beta = 0$ is called the **characteristic** equation of the recurrence (5).

If $\alpha^2 - 4\beta \geq 0$ (i.e., the characteristic equation has real roots), let $a, b$ be the roots. Then $\alpha = -a - b$ and $\beta = ab$, thus (5) becomes

$$u_{n+1}u_n - (a+b)u_n + ab = 0\,, \quad n \geq 0. \tag{6}$$

To this recurrence we will attach the sequence $(v_n)_{n\geq 0}$ defined by

$$v_{n+1}v_n - (a+b)v_{n+1} + ab = 0\,, \quad n \geq 0. \tag{7}$$

The sequences $(u_n)_{n\geq 0}$, $(v_n)_{n\geq 0}$ will be called **dual sequences**. This concept is perfectly motivated, as the dual of the dual of $(u_n)$ is exactly $(u_n)$. To ease the study of the sequences defined by (6) and (7) we will suppose that these recurrences are nontrivial, i.e.,

$$a, b \in \mathbb{R}^*, \tag{8}$$

and denote $c = \dfrac{b}{a}$. We will need the following

**Lemma.** *If the sequence $(y_n)$ verifies the recurrence relation*

$$y_{n+1} = cy_n + d, \quad n \geq 0 \quad (c \neq 0),$$

*then*

$$y_{n+p} = c^p y_p + d \sum_{k=p}^{n+p-1} c^{n+p-k}, \quad n, p \in \mathbb{N}^*.$$

**Proof.** We can equivalently write the recurrence relation as

$$\frac{y_{n+1}}{c^{n+1}} = \frac{y_n}{c^n} + \frac{d}{c^{n+1}},$$

which we can simplify as follows: we write this relation for $n$ taking in turn the values $p, p+1, \ldots, p+n-1$, we sum up the relations side by side and after all cancellations we get exactly the identity in the lemma. □

We can formulate now the first important result of this section.

**Theorem 1.** *If the sequence* $(u_n)_{n\geq 0}$ *verifies* (6) *and* (8), *then*

1) *For* $a \neq b$ *we have*

$$u_{n+p} = \frac{a(u_p - b) + b(a - u_p)c^n}{u_p - b + (a - u_p)c^n}, \quad \forall n, p \in \mathbb{N}; \tag{9}$$

2) *For* $a = b$ *we have*

$$u_{n+p} = \frac{a(n+1)u_p - na^2}{nu_p - (n-1)a}, \quad \forall n, p \in \mathbb{N}. \tag{9'}$$

**Proof.** Let us assume there is a rank $m \in \mathbb{N}$ so that

$$u_m = b$$

and the sequence $(u_n)$ verifies (6) and (8). Then we necessarily have

$$u_n = b, \text{ for any } n \in \mathbb{N}. \tag{10}$$

Indeed if $u_m = b$, then from (6) we get

$$u_{m-1} = \frac{ab}{a + b - u_n} = \frac{ab}{a} = b,$$

hence $u_{m-2} = b$ and so on, $u_n = b$, for any $n \in \mathbb{N}, n < m$.

For $n > m$, the relation (10) can be proved by induction. In this case it can be easily proven that (9) and (9′) are true.

Supposing now that $u_n \neq b$, for any $n \in \mathbb{N}$ , we can use the substitution

$$y_n = \frac{1}{u_n - b}, \quad n \geq 0 \tag{11}$$

which, considering (6) and the fact that (8) implies $u_n \neq a + b$, $\forall n \geq 1$, gives

$$y_n = \frac{1}{u_n - b} = \frac{1}{\dfrac{ab}{a + b - u_n} - b} = -\frac{1}{b} \cdot \frac{u_{n+1} - b - a}{u_{n+1} - b},$$

which is a linear, inhomogeneous recurrence. Using the lemma we get

$$y_{n+p} = c^p y_p + \frac{1}{b} c^{n+p} \sum_{k=p}^{n+p-1} \frac{1}{c^k}, \quad n, p \in \mathbb{N}^*. \tag{12}$$

If $c \neq 1$, i.e., $a \neq b$, then by computing the sum in (12) we get

$$y_{n+p} = c^p y_p + \frac{1 - c^n}{a - b}, \quad n, p \in \mathbb{N}^*,$$

and, considering (11), we obtain

$$\frac{1}{u_{n+p}-b}=\frac{u_p-b+(a-u_p)c^n}{(a-b)(u_p-b)}, \quad n,p\in\mathbb{N}^*. \tag{12'}$$

The denominators of the fractions (12′) are, in our conditions, both nonzero.

Because the numerator in the left-hand side is not zero, the one in the right-hand side must be different from zero as well. On the other hand, because none of the transformations used to get (12′) involve any indetermination, from (12′) we automatically get (9).

In the other case, when $c=1$, i.e., $a=b$, (12) gives

$$y_{n+p}=y_p+\frac{n}{b}, \quad n,p\in\mathbb{N}^*$$

and, having in mind the substitution we used, we get

$$\frac{1}{u_{n+p}-b}=\frac{nu_p-(n-1)b}{b(u_p-b)}, \quad n,p\in\mathbb{N}^*. \tag{12''}$$

As in the previous case, we can say that because $(u_n)_{n\geq 0}$ verifies (6) and (8), the operations leading from (12″) to (9′) do not involve any indetermination. □

**Remarks.** 1) In the proof of Theorem 1 we can use the alternative substitution

$$y_n=\frac{1}{u_n-a},$$

and then follow similar steps, because (6) and (7) are symmetrical with respect to $a$ and $b$.

2) If we take $p=0$ in Theorem 1, we get exactly the formula of the general term of the sequence $(u_n)_{n\geq 0}$.

Now we let the readers solve by themselves

**Problem 13d.** *Prove that if* $(u_n)_{n\geq 0}$ *verifies* (6) *and* (8), *then*

$$u_n=\begin{cases}\dfrac{a(u_0-b)+b(a-u_0)\left(\frac{b}{a}\right)^n}{u_0-b+(a-u_0)\left(\frac{b}{a}\right)^n}, & \text{if } a\neq b,\\[2ex] \dfrac{b(n+1)u_0-nb^2}{nu_0-(n-1)b}, & \text{if } a=b,\end{cases}$$

*for any* $n\in\mathbb{N}$.

We also let them prove the similar result for the dual sequence from

**Theorem 2.** *If the sequence* $(v_n)_{n\geq 0}$ *verifies* (7) *and* (8), *then*

1) *For* $a\neq b$,

$$v_{n+p}=\frac{b(v_p-a)+a(b-v_p)c^n}{v_p-a+(b-v_p)c^n}, \quad n,p\in\mathbb{N}. \tag{13}$$

2) *For* $a = b$,

$$v_{n+p} = \frac{na^2 - (n-1)av_p}{(n+1)a - v_p}, \quad n, p \in \mathbb{N}. \tag{13'}$$

**Remark.** If in Theorem 2 we take $p = 0$ we get the formula for the general term of the sequence $(v_n)_{n\geq 0}$. We therefore invite the readers to solve

**Problem 13e.** *Show that if* $(v_n)_{n\geq 0}$ *satisfies* (7) *and* (8), *then for any* $n \in \mathbb{N}$

$$v_n = \begin{cases} \dfrac{b(v_0 - a) + a(b - v_0)\left(\frac{b}{a}\right)^n}{v_0 - a + (b - v_0)\left(\frac{b}{a}\right)^n}, & \text{if } a \neq b, \\ \dfrac{na^2 - (n-1)av_0}{(n+1)a - v_0}, & \text{if } a = b. \end{cases}$$

Now using Theorem 1 and Theorem 2 we can prove a result regarding the convergence of the dual homographic sequences.

**Theorem 3.** *The dual sequences* $(u_n)_{n\geq 0}$, $(v_n)_{n\geq 0}$ *defined by* (6) *and* (8), *and* (7) *and* (8), *respectively, are both convergent and*

$$\lim_{n\to\infty} u_n, \ \lim_{n\to\infty} v_n \in \{a, b\}.$$

**Proof.** Indeed, for $a = b$ we have $\lim\limits_{n\to\infty} u_n = \lim\limits_{n\to\infty} v_n = a$. For $a \neq b$, we have

$$\lim_{n\to\infty} u_n = \begin{cases} a, & \text{if } |a| > |b|, \\ b, & \text{if } |a| < |b|. \end{cases}$$

We will now study the set of values which are not admitted for a dual homographic sequence. A number $\alpha \in \mathbb{R}$ will be called a *not admitted value* for a homographic sequence if it does not belong to the sequence, i.e., the existence of a certain $m \in \mathbb{N}$ so that $u_m = \alpha$ ($v_m = \alpha$) would imply that there is a certain rank $k$ (depending on $m$) for which the recurrence relation (6) or (7), respectively, does not hold. □

**Theorem 4.** *The set of not admitted values for a sequence* $(u_n)_{n\geq 0}$ *defined by* (6) *and* (8) *is formed by the terms of its dual* $(v_n)_{n\geq 0}$, *defined by* (7) *and* (8) *and the initial condition* $v_0 = 0$.

**Proof.** For $a \neq b$, using (9), the arguments in the proof of Theorem 1 and the fact that the denominator of the fraction in (9) must be different from zero, we get

$$u_p \neq \frac{ac^n - b}{c^n - 1}, \quad p \in \mathbb{N}, \ n \in \mathbb{N}^*. \tag{14}$$

But the right-hand side in (14) is exactly the term of rank $n-1$ in the sequence $(v_n)_{n\geq 0}$, given by (7), (8) and the initial condition $v_0 = 0$, which can be verified by using (13). For $a = b$ we follow the same steps, on the condition that the denominator of the fraction in (9′) is different from zero. We get

$$u_p \neq \frac{n-1}{n} \cdot b, \quad p \in \mathbb{N}, \ n \in \mathbb{N}^*,$$

where $\frac{n-1}{n}\cdot b$ is exactly the term of rank $n-1$ in the sequence $(v_n)_{n\geq 0}$ obtained with (7), (8) and the initial condition $v_0 = 0$, which results easily from (13′). □

In a similar manner we can prove

**Theorem 5.** *The set of not admitted values for the sequence $(v_n)_{n\geq 0}$ defined by (7) and (8) is formed by the terms of its dual $(u_n)_{n\geq 0}$, defined by (6), (8) and the initial condition $u_0 = a + b$.*

**Remark.** As the previous examples have shown, exploring around the source problem leads us to the discovery of a real "old mine", a huge amount of new problems which cannot wait to be formulated. We will write now only some of them, letting the readers solve them completely and discover by themselves more problems of the same family.

**Problem 13f.** *Show that if the sequence $(x_n)$ satisfies*

$$x_{n+1}x_n - 2x_n = 3, \quad n \geq 0,$$

*then none of its terms can belong to the set*

$$E = \left\{ \frac{3(-1)^{n+1} + 3^{n+1}}{(-1)^{n+1} - 3^{n+1}} \;\middle|\; n \in \mathbb{N} \right\}.$$

**Hint.** Here $a = 3$, $b = -1$ and Theorem 4 gives the conclusion.

**Problem 13g.** *Show that the sequence $(x_n)$ defined by*

$$x_{n+1}x_n - 2x_n = 3, \quad n \geq 0$$

*is convergent for any $x_0$ that does not belong to the set*

$$E = \left\{ \frac{3(-1)^{n+1} + 3^{n+1}}{(-1)^{n+1} - 3^{n+1}} \;\middle|\; n \in \mathbb{N} \right\}.$$

The set of not admitted values for the sequences in Problems 13f and 13g contains, of course, the values we considered in Problems 13, 13b and 13c, i.e., 0, $-\frac{3}{2}$, $-\frac{6}{7}$ etc. As we have already seen there, the set of not admitted values is the union

$$\{0\} \cup \left[-\frac{3}{2}, -\frac{6}{7}\right],$$

hence from the previous considerations we obtain the following new problem.

**Problem 13h.** *Show that for any $n \in \mathbb{N}$ we have*

$$-\frac{3}{2} \leq \frac{3(-1)^{n+1} + 3^{n+1}}{(-1)^{n+1} - 3^{n+1}} \leq -\frac{6}{7}.$$

It is also interesting to notice the form of the numbers $e_n$ from the set $E$ of numbers in Problem 13h. They represent, in an irreducible form, fractions which have, for denominator and numerator, successive natural numbers, alternately:

| $n$ | 0 | 1 | 2 | 3 | 4 |
|---|---|---|---|---|---|
| $e_n$ | 0 | $-\frac{3}{2}$ | $-\frac{6}{7}$ | $-\frac{21}{20}$ | $-\frac{60}{61}$. |

Therefore we can formulate the following two new problems.

**Problem 13i.** *Find the formula for the general term of the sequence*

$$\frac{3}{2}, \frac{6}{7}, \frac{21}{20}, \frac{60}{61}, \dots .$$

**Problem 13j.** *Show that the sequence in the previous problem is convergent and has the limit* 1.

All these problems were discovered starting from a particular recurrence relation, that in Problem 13. So the readers can obtain similar interesting problems, which are rather difficult to solve, by using other methods than those presented here. The considerations in this paragraph have led us to a very special conclusion: when talking about a recurrence as in (5), we do not need to impose any **initial condition**. This condition is verified implicitly, being intrinsic to the recurrence (5). Therefore a problem of the type 13, 13a could be simply formulated as follows.

**Problem 13′.** *Prove that any sequence of real numbers* $(x_n)_{n\geq 0}$ *which verifies*

$$x_{n+1}x_n - 2x_n = 3, \quad n \geq 0$$

*is convergent.*

We have to remind the reader that all the previous results were obtained in the case when the characteristic equation had **real** roots. Let us see now what happens in the other case.

## 13.5 The Characteristic Equation Has Complex Roots

In this case we have $\alpha^2 - 4\beta < 0$ and following the same steps as before we conclude that the results that still hold are Theorems 1, 2 and 4, but not Theorem 3. This last one has now a completely different statement.

First, we notice that in this case the numbers $a$ and $b$ for which

$$\alpha = -a - b \text{ and } \beta = ab$$

are conjugate complex numbers, i.e.,

$$a, b \in \mathbb{C}^* \setminus \mathbb{R} \text{ and } \bar{a} = b. \tag{8'}$$

**Theorem 6.** *If there exists* $m \in \mathbb{N}^*$ *and* $k \in \{1, 2, \dots, 2m-1\}$ *such that*

$$\arg a = \frac{k\pi}{m},$$

*then the dual sequences* $(u_n)_{n\geq 0}, (v_n)_{n\geq 0}$ *defined by* (6), (8′) *and* (7), (8′), *respectively, are periodical of period* $m$.

**Proof.** We have

$$\frac{a}{a}=\frac{a}{\overline{a}}=\frac{a^2}{a\cdot\overline{a}}=\left(\frac{a}{|a|}\right)^2=\left(\cos\frac{k\pi}{m}+i\sin\frac{k\pi}{m}\right)^2$$

and therefore

$$\left(\frac{a}{b}\right)^m=\cos 2k\pi+i\sin 2k\pi=1.$$

Then

$$\left(\frac{a}{b}\right)^{n+m}=\left(\frac{a}{b}\right)^n,\ \text{for any } n\in\mathbb{N},$$

and using Theorem 1 and Theorem 2, respectively, we get that the sequences $(u_n)_{n\geq 0}$ and $(v_n)_{n\geq 0}$ are periodical, of period $m$. □

An immediate application of Theorem 6 is given by

**Problem 13k.** *Prove that the sequence* $(x_n)$ *given by*

$$x_{n+1}x_n-2x_{n+1}+2=0,\ \ n\geq 0$$

*is periodical.*

**Solution 1.** We have $\alpha=-2,\ \beta=2$ and so $a=1+i,\ b=1-i$. Because

$$\arg a=\frac{\pi}{4},$$

by applying Theorem 6 we get that $(x_n)_{n\geq 0}$ is periodical, of period $m=4$. □

**Solution 2.** First let us remark that we cannot have $x=2$ (this would lead to $2=0$). By induction, we prove that $x_n\neq 2,\ \forall\ n\in\mathbb{N}$, and so

$$x_{n+1}=\frac{2}{2-x_n},\quad n\geq 0.$$

Therefore

$$x_{n+2}=\frac{2}{2-\dfrac{2}{2-x_n}}=\frac{2-x_n}{1-x_n},$$

and then

$$x_{n+3}=\frac{2(1-x_n)}{-x_n}.$$

Finally we get $x_{n+4}=x_n$. □

By carefully examining the second method, we see that the only not admitted value for this sequence is $\{2\}$. This situation differs essentially from that in Problems 13-13j, where the sets of not admitted values are infinite, but countable.

If we apply Theorem 5, we get the same result. Indeed, the dual of $(x_n)$ in Problem 13k is $(y_n)$, defined by the recurrence

$$y_{n+1}y_n-2y_n+2=0,\quad n\geq 0 \tag{15}$$

and, according to Theorem 5, the values not admitted for $(x_n)$ are given by the sequence $(y_n)$ with the initial condition $y_0 = a + b$. We have $a = 1 + i$, $b = 1 - i$ and thus $y_0 = 2$. From (15) we have, by induction, $y_n = 2, \forall n \geq 2$ and so, indeed, $\{2\}$ is the only not admitted value for the sequence $(x_n)$.

Therefore, a homographic sequence which is convergent implies an infinite set of not admitted values for that sequence, while the periodicity of such a sequence is associated with the existence of only one not admitted value for it.

We can reformulate this result as

**Problem 13l.** *Prove that the sequence $(x_n)_{n\geq 0}$ defined by*

$$x_{n+1}x_n + ax_n + b = 0, \quad n \geq 0 \tag{16}$$

*is periodical if and only if there is only one value $\alpha \in \mathbb{R}$ so that, taking $x_0 = \alpha$,* (16) *does not hold for all values of $n \in \mathbb{N}$.*

## 13.6 Homographic Functional Equations

A sequence is a particular function, which is defined on $\mathbb{N}$, the set of positive integer numbers. Therefore we can exploit the following analogy between homographic recurrences and functional homographic equations:

$$f(n+1) = \frac{af(n)+b}{cf(n)+d}, \quad n \in \mathbb{N} \quad | \quad f(x+1) = \frac{af(x)+b}{cf(x)+d}.$$

Using only a part of the results obtained previously for homographic recurrences, we can formulate new problems about functional equations. An example is

**Problem 13m.** *Prove that the function $f : \mathbb{R} \to \mathbb{R}\backslash\{2\}$ which verifies*

$$f(x+1) \cdot f(x) - 2f(x+1) + 2 = 0, \quad \forall\, x \in \mathbb{R}$$

*is periodical.*

**Solution.** The same as for Problem 13k, the only difference being that here we have a function defined on $\mathbb{R}$. As $f(x) \neq 2$, $\forall\, x \in \mathbb{R}$, we have

$$f(x+1) = \frac{2}{2-f(x)}, \quad x \in \mathbb{R},$$

i.e., $f(x+4) = f(x)$, $\forall\, x \in \mathbb{R}$, so $f$ is periodical of period $T = 4$. □

The text of the previous problem can be strengthened as follows.

**Problem 13n.** *Prove that the function $f : \mathbb{R} \to \mathbb{R}\backslash\{2\}$ which verifies*

$$f(x+1) \cdot f(x) - 2f(x) + 2 = 0, \quad \forall\, x \in \mathbb{R}$$

*is periodical.*

**Solution.** If there exists $x_0 \in \mathbb{R}$ for which $f(x_0) = 2$, then $f(x_0 + k) = 2, \forall\, x \in \mathbb{R}$. If there exists $x_1 \in \mathbb{R}$ so that $f(x_1) \neq 2$, then we prove similarly that $f(x_1 + k) \neq 2, \forall\, k \in \mathbb{N}$ and $f(x_1 + 4) = f(x_1)$.

These two cases show that, for any $x \in \mathbb{R}$, $f(x+4) = f(x)$, which finishes the solution. □

## 13.7 Bibliographical Comments

Problem 13 was proposed by D.M. Batinetu-Giurgiu at the County Round of the National Mathematics Olympiad in 1984 [7]. A problem of the same kind was proposed at the W.L. Putnam Mathematics Competition (see [6]).

Problem 13a, which generalizes Problem 13, is in fact the problem around which we have written our work [2]. All the three methods of solving Problem 13 are particular cases of the solutions of Problem 13a given in [2].

Problem 13c is in fact Application 2, p. 19 in [3], a paper where we stressed the convergence criteria for the sequences studied there, a topic which is also found in Section 22 of this book.

The paragraphs 13.3–13.5 are taken with insignificant changes from our paper [4]. However in paragraph 13.4 we have added some applications that are effectively tackled with the available methods.

Problem 13k is an extension of Problem 2569 in *Revista Matematica a Elevilor din Timisoara*, no. 2/1976, author Titu Andreescu, that was also published in *Gazeta Matematica* no. 10/1980 as problem 18474*.

## 13.8 References


1. **C. Avadanei et al.,** *From Elementary Mathematics to Higher Mathematics.* (Romanian), Editura Academiei, 1987

2. **V. Berinde,** *On a contest problem.* (Romanian) Gazeta Matematica, **90** (1985), no. 6, 186–188

3. **V. Berinde,** *On a Necessary and Sufficient Criterion for the Convergence of Sequences.* (Romanian), ASTRA Matematica, **1** (1990), No. 5, 18–21

4. **V. Berinde,** *The Duality in the Study of Homographical Recurrences.* (Romanian) Lucr. Semin. Creativ. Mat., **1** (1991-1992), 1–14

5. **M. Cocuz,** *Mathematical Problems.* (Romanian), Editura Academiei, 1984

6. **A.M. Gleason,** *The W.L. Putnam Mathematical Competition,* The Mathematical Association of America, 1980

7. **I. Tomescu et al.,** *Problems in the Mathematical Olympiad for Secondary Schools (1950-1990).* (Romanian), Editura Stiintifica, 1992

# Theme 14

# Exponential Equations Having Exactly Two Solutions

## 14.1 Source Problem

The starting point of all investigations and explorations in this section is

**Problem 14.** *Solve the equation*

$$4^x 9^{\frac{1}{x}} + 4^{\frac{1}{x}} 9^x = 210. \tag{1}$$

**Solution.** The condition that has to be satisfied is $x \in \mathbb{R}^*$. The given equation is then equivalent to

$$2^3\left[\left(2^2 \cdot 3^{-\frac{1}{x}}\right)^{x-2} - 1\right] + 3^3\left[\left(2^{-\frac{1}{x}} \cdot 3^2\right)^{x-2} - 1\right] = 0. \tag{1'}$$

**Case I. x < 0.** In this case we have $-\dfrac{1}{x} > 0$, so

$$2^{-\frac{1}{x}} > 1 \text{ and } 3^{-\frac{1}{x}} > 1$$

which shows that

$$2^2 \cdot 3^{-\frac{1}{x}} > 1 \text{ and } 2^{-\frac{1}{x}} \cdot 3^2 > 1, \tag{2}$$

i.e.,

$$\left(2^2 \cdot 3^{-\frac{1}{x}}\right)^{x-2} < 1 \text{ and } \left(2^{-\frac{1}{x}} \cdot 3^2\right)^{x-2} < 1,$$

since $x - 2 < 0$. This shows that (1′) cannot hold for $x < 0$.

**Case II. x > 0.** Here we consider first the situation when $x > 1$. This implies $-\dfrac{1}{x} > -1$ and so

$$2^{-\frac{1}{x}} \geq \frac{1}{2} \text{ and } 3^{-\frac{1}{x}} \geq \frac{1}{3},$$

which shows that the relations (2) hold in this case as well.

We notice that for $x > 2$ the square brackets in $(1')$ are strictly positive, while for $1 < x < 2$ the same brackets are strictly negative. So we can have equality in $(1')$ if and only if both brackets are equal to zero, which is accomplished for $x = 2$.

This proves that $x_1 = 2$ is the only solution of the equation (1) in the interval $(1, \infty)$. We now denote

$$f(x) = 4^x 9^{\frac{1}{x}} + 4^{\frac{1}{x}} 9^x, \quad x \in \mathbb{R}^*.$$

It is easy to notice that $f\left(\dfrac{1}{x}\right) = f(x)$, for any $x \in \mathbb{R}^*$. As $t(x) = \dfrac{1}{x}$ is a bijection of $(0,1)$ onto $(1,\infty)$, we deduce that $x_2 = \dfrac{1}{2}$ is the only solution of the equation (1) in the interval $(0,1)$.

So the given equation has the solutions 2 and $\dfrac{1}{2}$. □

**Remarks.** $1^0$ The two solutions of the equation (1) satisfy

$$x_1 x_2 = 1. \tag{3}$$

$2^0$ Analyzing the way in which we have solved the source problem, the algorithmicity and the generality of the method we used come out. This leads us to the idea that, by replacing the numbers 4 and 9 with two constants $a$ and $b$ having a certain relationship and by changing the right side in an adequate way, we could obtain an equation with similar properties. Essentially we are interested in obtaining an equation with exactly two solutions $x_1$ and $x_2$ that satisfy (3).

In order to start the investigations, one has to notice that (1) is equivalent to

$$2^{2x} \cdot 3^{\frac{2}{x}} + 2^{\frac{2}{x}} \cdot 3^{2x} = 2^{2^2} \cdot 3 + 2 \cdot 3^{2^2},$$

which suggests that we study the solvability of the equation

$$a^{cx} \cdot b^{\frac{c}{x}} + a^{\frac{c}{x}} \cdot b^{cx} = a^{c^2} b + ab^{c^2}, \tag{4}$$

where $a, b, c \in (0,\infty)\backslash\{1\}$ and $c \in \mathbb{R}^*$.

It is obvious that (1) is obtained from (4) when $a = 2,\ b = 3$ and $c = 2$.

## 14.2 Exploring the Solvability of the General Equation

Following the same steps as in the case of the source problem, we will first rewrite (4) in the equivalent form

$$a^{c^2-1}\left[\left(a^c b^{-\frac{1}{x}}\right)^{x-c}-1\right]+b^{c^2-1}\left[\left(b^c a^{-\frac{1}{x}}\right)^{x-c}-1\right]=0. \tag{4'}$$

This shows that, in order to obtain some information on the sign of the square brackets in (4′), we have to get some relationships between $a^c$ and $b$, and between $b^c$ and $a$, respectively.

Thus we obtain the following result.

**Lemma 1.** *The following statements are equivalent:*

(i) $a, b \in \mathbb{R}_+^*$ *and there exists* $c > 1$ *so that*

$$a^c > b \text{ and } b^c > a; \tag{5}$$

(ii) $a, b \in (1, \infty)$.

**Proof.** The case when $a = b$ is trivial, so that we are going to analyze only the situation $a \neq b$. We consider, for example, $a < b$ (the case $a > b$ will be implied by symmetry).

We will first prove that (ii) $\Rightarrow$ (i).

As $b > 1$, for $c > 1$ we will obtain that $b^c > b > a$. On the other hand, when $a > 1,\ b > 1$ the equation $a^x = b$ has only one solution, that is $x_0 = \log_a b \in (1, \infty)$. So the relation $a^c > b$ holds for any $c > x_0$, and hence (5) is satisfied.

Conversely, we assume that (i) is true and we try to show that this implies (ii). We suppose that $a < 1$. In this case

$$a^c < a < b$$

which contradicts (5). So if (i) holds, then we must have $a > 1$, which implies $b > 1$ (since $a < b$).

For our proof to be complete, we suppose $b < 1$. In this case $a < 1$, too, so that from (5) we obtain

$$b^c > a > b^{\frac{1}{c}},$$

which contradicts $c > 1$. This shows that $b > 1$. The relation (5) implies that $a > b^{\frac{1}{c}}$, i.e., $a > 1$. □

**Lemma 2.** *The following statements are equivalent:*

(i) $a, b \in \mathbb{R}_+^*$ *and there exists* $c > 1$ *so that*

$$a^c < b \text{ and } b^c < a; \tag{6}$$

(ii) $a, b \in (0, 1)$.

**Proof.** We denote $a_1 = \frac{1}{a}$, $b_1 = \frac{1}{b}$ and use Lemma 1. □

Now we are able to solve

**Problem 14a.** *If* $a, b \in (0, \infty)$ *and* $c > 1$ *satisfy*

$$(a^c - b)(b^c - a) > 0, \tag{7}$$

*then equation* (4) *has exactly two solutions*

$$x_1 = c \textit{ and } x_2 = \frac{1}{c}.$$

**Solution.** We will have two distinct cases: 1) $a^c > b$; 2) $a^c < b$.
**Case $\mathbf{a^c > b}$.** Then condition (7) implies that $b^c > a$ and so, by Lemma 1, we obtain that $a, b \in (1, \infty)$. For each $x \in (1, \infty)$ we will have $\frac{1}{x} < 1$ and so

$$b > b^{\frac{1}{x}} \text{ and } a > a^{\frac{1}{x}}. \tag{8}$$

Taking into account that $a^c > b$ and $b^c > a$, we will obtain that

$$a^c > b^{\frac{1}{x}} \text{ and } b^c > a^{\frac{1}{x}},$$

i.e.,

$$a^c \cdot b^{-\frac{1}{x}} > 1 \text{ and } b^c \cdot a^{-\frac{1}{x}} > 1. \tag{9}$$

This shows that, for $x \in (1, \infty)$, we can have equality in (4′) if and only if both square brackets are equal to zero, in the other cases the left side being strictly positive (respectively strictly negative).

The equality to zero holds for $x = c$. So $x_1 = c$ is a solution of the equation (4) on $(1, \infty)$.

Let us now consider the case $x \in (0, 1)$. We notice that

$$f(x) = a^{cx} b^{\frac{c}{x}} + a^{\frac{c}{x}} b^{cx}, \quad x \in \mathbb{R}^* \tag{*}$$

has the property

$$f\left(\frac{1}{x}\right) = f(x), \quad \forall x \in \mathbb{R}^*.$$

As $t(x) = \frac{1}{x}$ is a bijection of $(0, 1)$ onto $(1, \infty)$, we deduce that $\frac{1}{x} = c$ is the only solution in $(0, 1)$ of the equation (4).

If we consider now the case $x \in (-\infty, 0)$, we notice that (8) and (9) do still hold and, as $x - c < 0$, the left side of (4′) will be strictly negative. So there is no solution on $(-\infty, 0)$.

As a conclusion, when $a^c > b$ the equation (4) has only two solutions, i.e., $x_1 = c$ and $x_2 = \frac{1}{c}$.

**Case $\mathbf{a^c < b}$.** In this situation, (7) implies $b^c < a$ and, by applying Lemma 2, we obtain that $a, b \in (0,1)$. Further on we follow the same steps as above. □

**Remark.** In the source problem we have the assumption $a, b \in (1, \infty)$. But as one can see above, we managed to solve the equation also in the case $a, b \in (0,1)$.

And now we invite readers to solve the following problem.

**Problem 14b.** *Solve the equation*

$$4^{-x}9^{-\frac{1}{x}} + 4^{-\frac{1}{x}}9^{-x} = \frac{35}{1296}.$$

**Hint.** The equation can be written as

$$\left(\frac{1}{2}\right)^{2x} \cdot \left(\frac{1}{3}\right)^{\frac{2}{x}} + \left(\frac{1}{2}\right)^{\frac{2}{x}} \cdot \left(\frac{1}{3}\right)^{2x} = \frac{35}{1296},$$

hence we have here $a = \frac{1}{2}$, $b = \frac{1}{3}$ and $c = 2$ and then apply Lemma 2. □

Similar results to those obtained above by exploring Problem 14 could have been obtained if we had started from any of the equations below, which are obviously related to the one initially considered:

$$4^{x}9^{-\frac{1}{x}} + 4^{-\frac{1}{x}}9^{x} = \frac{275}{6}; \tag{10}$$

$$\left(4^{x} + 9^{\frac{1}{x}}\right)\left(4^{\frac{1}{x}} + 9^{x}\right) = 1577; \tag{11}$$

$$(4^{x} + 9^{x})\left(4^{\frac{1}{x}} + 9^{\frac{1}{x}}\right) = 485; \tag{12}$$

$$(4^{x} + 9^{x})\left(4^{-\frac{1}{x}} + 9^{-\frac{1}{x}}\right) = \frac{485}{6}. \tag{13}$$

We recommend that readers study them thoroughly in order to obtain more general results.

## 14.3 Other Investigation Directions

However pleased we are with the results obtained in the previous paragraphs, results that can be well applied also to the equations (10)–(13), we will not end this section without answering the natural question: are the characterizations in Lemma 1 and Lemma 2 specific only to **pairs** of numbers or can they be extended to a system of $n$ numbers, where $n \geq 3$?

Following the proofs of the two lemmas mentioned above, we are easily led to

**Lemma 3.** *The following statements are equivalent:*

(i) $a_1, a_2, \ldots, a_n \in (1, \infty)$;

(ii) $a_1, a_2, \ldots, a_n \in \mathbb{R}_+^*$ *and there exists* $c > 1$ *so that* $a_1^c > a_2$, $a_2^c > a_3, \ldots$, $a_{n-1}^c > a_n$, $a_n^c > a_1$.

**Lemma 4.** *The following statements are equivalent:*
(i) $a_1, a_2, \ldots, a_n \in (0,1)$;
(ii) $a_1, a_2, \ldots, a_n \in \mathbb{R}_+^*$ *and there exists* $c > 1$ *so that* $a_1^c < a_2$, $a_2^c < a_3, \ldots,$ $a_{n-1}^c < a_n$, $a_n^c < a_1$.

Using these two lemmas we will be able now to solve the following generalization of Problem 14a.

**Problem 14c.** *Solve the equation*

$$a_1^{cx-1}a_2^{\frac{c}{x}-1} + \ldots + a_{n-1}^{cx-1}a_n^{\frac{c}{x}-1} + a_n^{cx-1}a_1^{\frac{c}{x}-1} = a_1^{c^2-1} + \ldots + a_n^{c^2-1},$$

*knowing that* $c > 1$ *and* $a_1, a_2, \ldots, a_n \in \mathbb{R}_+^*$ *so that*

$$a_1^c - a_2,\ a_2^c - a_3, \ldots, a_{n-1}^c - a_n,\ a_n^c - a_1$$

*have the same sign.*

A particular case would be more encouraging here.

**Problem 14d.** *Solve the equation*

$$2^{2x-1}3^{\frac{2}{x}-1} + 3^{2x-1}4^{\frac{2}{x}-1} + 2^{4x+\frac{2}{x}-3} = 99.$$

**Hint.** We are now in the case $n = 3$, $a_1 = 2$, $a_2 = 3$, $a_3 = 4$, $c = 2$. The equation has only the solutions $x_1 = 2$ and $x_2 = \frac{1}{2}$. □

**Remark.** It is easy to show that the following reciprocal of Problem 14a is also true.

If $x = c$ and $x = \frac{1}{c}$ are the only solutions of (4) and $c > 0$, then $a, b$ and $c$ do satisfy (7).

Indeed, the function $f$ defined by (*) is continuous on $\mathbb{R}^*$. As

$$f(c) = 0,\ f\left(\frac{1}{c}\right) = 0$$

and $f$ has no other zeroes between $\frac{1}{c}$ and $c$, $f(x)$ will have a constant sign for $x$ ranging between these values. This implies, considering the equivalence of (4) and (4′), that for any $x$ between $\frac{1}{c}$ and $c$ the numbers

$$b^c a^{-\frac{1}{x}} - 1 \text{ and } a^c b^{-\frac{1}{x}} - 1$$

will have the same sign. As $x = 1$ lies certainly between $c$ and $\frac{1}{c}$, (7) is implied.

## 14.4 Bibliographical Comments

We do not know the very origin of Problem 14. The oldest bibliographical source we could find in Romanian literature is [2], where the problem appears without a solution. The one who has called my attention on this problem, in the years 1980–1981, is Mr. Gh. Onisa, a mathematics teacher at the National College "Vasile Lucaciu" Baia Mare, whom I want to thank this way, too.

Lemma 1, discovered by the author when investigating around Problem 14, was published in *Gazeta Matematica* no. 5/1989 as the proposed problem 21776.

Problem 14a is an extension to the case $a, b \in (0,1)$ of the Problem C:117 published in *Gazeta Matematica* no. 5/1981 and having Gh. Szollosi as author.

Problems 10–12 were published in *Gazeta Matematica* as follows: the problem 19911, no. 10–11/1983 (author M. Ghita); the problem 19984, no. 1/1984 (author M.O. Drambe); respectively the problem 20516, no. 8/1985 (author V. Berinde). Problem 13 was proposed in [1].

## 14.5 References


1. **V. Berinde,** *Exponential Equations with Exactly Two Solutions.* (Romanian), Lucr. Semin. Creativ. Mat., **4** (1994–1995), 37–44

2. **V. Chiriac, M. Chiriac,** *Problems of Algebra.* (Romanian), Editura Tehnica, 1977

3. **O. Konnerth,** *On the Equation* $a^x b^{\frac{1}{x}} + a^{\frac{1}{x}} b^x = c$. (Romanian), Gazeta Matematica, **87** (1982), no. 2, 55–56

# Theme 15

# A Class of Functional Equations

As we have already mentioned in Section 13, there is a natural relationship between recurrent sequences and functional equations. Given a relationship between some successive terms of a sequence, we can determine the properties or even the general term of that sequence. The same happens in the case of functional equations: a relation characterizing a function can lead us to the properties of the solutions or even to finding the solutions of this equation.

The best known functional equation is the classical Cauchy's equation. Many other functional equations can be reduced to this one. As the readers well know, the solutions of such an equation are sought in a certain class of functions as, for instance, the class of continuous functions (at a point or on a set), the class of functions with a (finite) limit at a point, the class of differentiable functions etc.

In other situations the properties of the solutions are intrinsic. In those cases there is no need of an a priori property of the solution — the very fact that the function is a solution of that functional equation implies its having the property in question. In many situations a pointwise property of the solution can lead to the corresponding global property, as we will see in the following.

## 15.1 Source Problem

The investigations of this section will start from the following

**Problem 15.** *Find the continuous functions $f : \mathbb{R} \to \mathbb{R}$ which satisfy*

$$f(2^x) = f(3^x), \ \forall \, x \in \mathbb{R}.$$

**Solution.** We denote $y = 2^x$ which implies $x = \log_2 y$ and so $3^x = 3^{\log_2 y} = 3^{\log_3 y / \log_3 2} = \left(3^{\log_3 y}\right)^{\log_2 3} = y^{\log_2 3}$. Now by denoting $\alpha = \log_2 3$ we will obtain $\alpha > 1$ and $3^x = y^\alpha$. As $y = 2^x > 0, \forall \, x \in \mathbb{R}$, the given relation can now be written as

$$f(y^\alpha) = f(y), \ \ y \in (0, \infty). \tag{1}$$

In (1) we replace $y$ by $y^{\frac{1}{\alpha}}$, then we do the same in the obtained relation and so on. We will have that

$$f(y) = f\left(y^{\frac{1}{\alpha}}\right) = f\left(y^{\frac{1}{\alpha^2}}\right) = \ldots = f\left(y^{\frac{1}{\alpha^n}}\right)$$

and so

$$f(y) = f\left(y^{\frac{1}{\alpha^n}}\right), \quad \forall\, y \in (0,\infty) \text{ and } \forall\, n \in \mathbb{N}. \tag{1'}$$

As $f$ is continuous and $\dfrac{1}{\alpha^n} \to 0$ when $n \to \infty$, we obtain that

$$\lim_{n\to\infty} f\left(y^{\frac{1}{\alpha^n}}\right) = f\left(\lim_{n\to\infty} y^{\frac{1}{\alpha^n}}\right) = f(1)$$

and hence from (1′) we deduce that $f(y) = f(1)$, $\forall\, y \in (0,\infty)$.

As $f$ is continuous, by denoting $f(1) = c$ we obtain that for any function $g : (-\infty, 0] \to \mathbb{R}$, continuous at zero and with $g(0) = c$, the function

$$f(x) = \begin{cases} g(x), & \text{for } x \in (-\infty, 0], \\ c, & \text{for } x > 0 \end{cases}$$

is a solution of the given functional equation.

## 15.2 A Close Examination of the Problem and its Solution. First Explorations

It is easy to notice that in the most important part of our solution we only used the continuity of $f$ at the point $x = 1$. So the next problem will have a similar solution.

**Problem 15a.** *Find all functions $f : \mathbb{R} \to \mathbb{R}$ which are continuous at $x = 1$ and satisfy*

$$f(2^x) = f(3^x), \ \forall\, x \in \mathbb{R}.$$

**Solution.** As $\alpha > 1$ and so $\dfrac{1}{\alpha^n} \to 0$, i.e., $y^{\frac{1}{\alpha^n}} \to 1$ when $n \to \infty$, $\forall y > 0$, (1′) will lead us, as before, to

$$f(y) = f(1), \ \forall\, y > 0.$$

But, since $f$ is **not continuous on** $\mathbb{R}$ in this case, $f$ will not be continuous at zero either. Denoting $c = f(1)$, the solution of the functional equation will be

$$f(x) = \begin{cases} c, & \text{for } x > 0, \\ g(x), & \text{for } x \le 0, \end{cases}$$

where $g : (-\infty, 0] \to \mathbb{R}$ is an arbitrary function. □

**Remark.** The continuity of $f$ at 1 has led us to the conclusion that $f$ is continuous on $(0, \infty)$. Therefore we can weaken this hypothesis further, as in

**Problem 15b.** *Find all functions* $f : \mathbb{R} \to \mathbb{R}$ *which have a finite limit at 1 and satisfy*

$$f(2^x) = f(3^x),\ \forall\, x \in \mathbb{R}.$$

**Solution.** In this case the existence of the finite limit at 1, be that limit $l$, implies that

$$\lim_{n\to\infty} f\left(y^{\frac{1}{a^n}}\right) = l,\ \text{for any } y > 0,\ y \neq 1,$$

so the solution of the functional equation will be

$$f(x) = \begin{cases} g(x)\,, & \text{for } x \le 0\,, \\ l\,, & \text{for } x \in (0,1) \cup (1,\infty)\,, \\ a & ,\ \text{for } x = 1 \end{cases}$$

where $g : (-\infty, 0] \to \mathbb{R}$ is an arbitrary function and $a \in \mathbb{R}$ is a constant. □

**Remark.** The attempt of our first explorations was to extend the class of functions among which we were seeking for the solution of the functional equation. If we denote

$$C = \{f : R \to R \mid f \ \text{ continuous}\};$$
$$C_1 = \{f : R \to R \mid f \ \text{ continuous at } \mathbf{1}\};$$
$$\text{and } L_1 = \{f : R \to R \mid f \ \text{ has a finite limit at } \mathbf{1}\},$$

then we easily notice that while extending Problem 15 we have exploited in fact the inclusions

$$C \subset C_1 \subset L_1.$$

There is also another direction to explore: extending the functional equation itself. Thus we obtain

**Problem 15c.** *Find all continuous functions* $f : \mathbb{R} \to \mathbb{R}$ *that satisfy*

$$f(\log_2 x) - f(\log_3 x) = \log_5 x,\ \ \forall x \in (0, \infty). \tag{2}$$

**Solution.** We denote $\log_2 x = y$ which implies $x = 2^y$. So

$$\log_3 x = \log_3 2^y = y \log_3 2 \text{ and } \ \log_5 x = y \log_5 2.$$

We also denote $a = \log_3 2$ and $b = \log_5 2$. Then $a, b \in (0,1)$ and so (2) becomes

$$f(y) - f(ay) = by, \quad y \in \mathbb{R}. \tag{2'}$$

We replace here $y$ by $ay$, then we do the same in the obtained relation and so on, till we get the following $n-1$ relations:

$$f(ay) - f(a^2y) = b \cdot ay,$$
$$f(a^2y) - f(a^3y) = b \cdot a^2y,$$
$$\dots\dots\dots\dots\dots\dots\dots\dots\dots\dots$$
$$f(a^{n-2}y) - f(a^{n-1}y) = b \cdot a^{n-2}y,$$
$$f(a^{n-1}y) - f(a^n y) = b \cdot a^{n-1}y, \quad \forall y \in \mathbb{R}.$$

Summing them and (2′) side by side and doing all cancellations we obtain

$$f(y) - f(a^n y) = \frac{b \cdot y(1 - a^n)}{1 - a}, \quad \forall n \in \mathbb{R}, \forall n \in \mathbb{N}^*. \tag{3}$$

As $f$ is continuous and $\lim_{n \to \infty} a^n = 0$, passing to the limit in (3) we have that

$$f(y) - f(0) = \frac{by}{1 - a}, \quad \forall y \in \mathbb{R},$$

which shows that the solution of the functional equation is

$$f(x) = \frac{\log_5 2}{1 - \log_3 2} \cdot x + f(0), \quad \forall x \in \mathbb{R},$$

where $f(0) = c$ is an arbitrary real number. □

Now taking into account the experience at Problem 15 we can formulate

**Problem 15d.** *Find all functions* $f : \mathbb{R} \to \mathbb{R}$ *which are continuous at zero and satisfy* (2).

**Solution.** We will obtain exactly the same solution as in the previous problem, for in solving the latter we have only used the hypothesis of Problem 15d. This result can be formulated as a new problem.

**Problem 15e.** *Show that if* $f : \mathbb{R} \to \mathbb{R}$ *satisfies* (2) *and is continuous at zero, then* $f$ *is continuous on* $\mathbb{R}$.

We can also weaken the assumptions in Problem 15d.

**Problem 15f.** *Find all functions* $f : \mathbb{R} \to \mathbb{R}$ *which have a finite limit at zero and satisfy* (2).

**Solution.** We denote

$$l = \lim_{y \to 0} f(y)$$

and from (3) we obtain now that

$$f(y) - l = \frac{by}{1 - a}, \quad \forall y \in \mathbb{R}^*.$$

So the solution of the functional equation will be

$$f(x) = \begin{cases} \dfrac{\log_5 2}{1 - \log_3 2} \cdot x + l\,, & \text{for } x \neq 0, \\ a\,, & \text{for } x = 0, \end{cases}$$

where $a$ is an arbitrary real constant. The solution is a continuous function on $\mathbb{R} \backslash \{0\}$. □

Using this last result we can also solve the following problem.

**Problem 15g.** *Prove that if $f : \mathbb{R} \to \mathbb{R}$ satisfies* (2) *and is continuous on the left (on the right) at zero, then $f$ is continuous on $\mathbb{R}$.*

A more complicated functional equation is involved in

**Problem 15h.** *Find all continuous functions $f : \mathbb{R} \to \mathbb{R}$ which satisfy*

$$f(2^x) - 2f(2^{x+1}) + f(2^{x+2}) = 2^{x+3}, \ \forall x \in \mathbb{R}. \tag{4}$$

**Solution.** We are going to study the more general problem when $2^x$ is replaced by $a^x$, $a > 1$. Denoting $y = a^x$ this becomes

$$f(y) - 2f(ay) + f(a^2y) = a^3y, \ \ y \in (0, \infty). \tag{4'}$$

We will first show how we get from this to a relation like (2′) and then we are going to follow the same steps as in solving Problem 15c. In (4′) we replace $y$ by $\frac{y}{a}$, then we do the same in the newly obtained relation and so on, till we get to

$$\begin{aligned}
&f\left(\frac{y}{a}\right) - 2f(y) + f(ay) = a^2y,\\
&f\left(\frac{y}{a^2}\right) - 2f(\frac{y}{a}) + f(y) = ay,\\
&f\left(\frac{y}{a^3}\right) - 2f(\frac{y}{a^2}) + f(\frac{y}{a}) = y,\\
&\dots\dots\dots\dots\dots\dots\dots\dots\\
&f\left(\frac{y}{a^{n+2}}\right) - 2f(\frac{y}{a^{n+1}}) + f(\frac{y}{a^n}) = \frac{y}{a^{n-1}}.
\end{aligned}$$

Summing side by side the last $n+1$ inequalities and cancelling like terms we obtain

$$f\left(\frac{y}{a^{n+2}}\right) - f\left(\frac{y}{a^{n+1}}\right) - f\left(\frac{y}{a}\right) + f(y) \tag{5}$$

$$= ay\left(1 + \frac{1}{a} + \frac{1}{a^2} + \dots + \frac{1}{a^n}\right), \forall y \in (0, \infty) \text{ and } \forall n \in \mathbb{N}^*.$$

But $a > 1$, so $\lim\limits_{n\to 0} \frac{y}{a^{n+1}} = \lim\limits_{n\to 0} \frac{y}{a^{n+2}} = 0, \forall y \in (0, \infty)$.

On the other hand, $f$ is continuous at zero, which implies

$$\lim_{n\to 0} f\left(\frac{y}{a^{n+1}}\right) = \lim_{n\to 0} f\left(\frac{y}{a^{n+2}}\right) = f(0).$$

Now passing to the limit in (5) we have that

$$f(0) - f(0) - f\left(\frac{y}{a}\right) + f(y) = \frac{a^2y}{a-1},$$

i.e.,

$$f(y) - f\left(\frac{y}{a}\right) = \frac{a^2y}{a-1}, \ y \in (0, \infty)$$

which is exactly a relation of the type (2′).

Using Problem 15c, we get that the solution of equation (4′) is

$$f(y) = \left(\frac{a}{1-a}\right)^2 \cdot ay + f(0), \ y \in (0, \infty)$$

and so the solution of Problem 15h will be the function

$$f(x) = \begin{cases} g(x), & \text{for } x < 0, \\ 8x + f(0), & \text{for } x \geq 0, \end{cases}$$

where $f(0)$ is an arbitrary real number and $g : (-\infty, 0] \to \mathbb{R}$ is a continuous function so that $g(0) = f(0)$. □

Now we invite the reader to study the following variants of Problem 15h and some other related problems.

**Problem 15i.** *Find all functions* $f : \mathbb{R} \to \mathbb{R}$ *which are continuous at zero and satisfy* (4).

**Problem 15j.** *Find all functions* $f : \mathbb{R} \to \mathbb{R}$ *which have a finite limit at zero and satisfy* (4).

**Problem 15k.** *Find all functions which are continuous at 1 and satisfy the functional equation*

$$f(4x-3) - 2f(2x-1) + f(x) = x - 1, \ \forall x \in \mathbb{R}.$$

**Problem 15l.** *Find all functions* $f : \mathbb{R} \to \mathbb{R}$ *which are continuous at* $x_0 = 1$ *and satisfy*

$$f(9x-8) - 2f(3x-2) + f(x) = 4x - 4, \ \forall x \in \mathbb{R}.$$

**Remark.** Each of the problems 15h–15l has three terms on the left-hand side. As the number of terms on the left side of such an equation can be reduced by one, using the method we described when solving Problem 15h, similar problems can be formulated starting from the following two.

**Problem 15m.** *Find all continuous functions* $f : \mathbb{R} \to \mathbb{R}$ *which satisfy*

$$f(2^x) - 3f(2^{x+1}) + 3f(2^{x+2}) - f(2^{x+3}) = 2^{x+4}, \ x \in \mathbb{R}.$$

**Problem 15n.** *Find all continuous functions* $f : \mathbb{R} \to \mathbb{R}$ *which satisfy for any* $x \in \mathbb{R}$ *(p is a positive integer,* $p \geq 1$*):*

$$C_p^0 f(2^x) - C_p^1 f(2^{x+1}) + C_p^2 f(2^{x+2}) - \ldots + (-1)^p C_p^p f(2^{x+p}) = 2^{x+p+1}.$$

## 15.3 Other Functional Equations. Non-Binomial Coefficients

One can easily notice that the functional equations considered so far contained the binomial coefficients $(1, -1)$, $(1, -2, 1)$, $(1, -3, 3, -1)$ etc. If we carefully evaluate the importance of these coefficients along the solution, we conclude that they can be replaced by other homogeneous combinations as, for instance, $(3, -4, 1)$, $(5, -3, -2)$ etc., so that the computations should not be essentially affected. Still, introducing such combinations implies some trouble that can be solved by keeping only some of the hypotheses we used in the problems above.

**Problem 15o.** *Prove that if* $f : \mathbb{R} \to \mathbb{R}$ *is continuous at zero,* $f(0) = 0$ *and*

$$3f(t) - 4f(4t) + f(16t) = 3t, \forall t \in \mathbb{R},$$

*then* $f$ *is continuous on* $\mathbb{R}$.

**Solution.** If we divide the given relation by $t \neq 0$ and denote

$$l = \lim_{t \to 0} \frac{f(t)}{t},$$

we obtain $l = 1$. As $f(0) = 0$ and $f$ is continuous at zero, we will have

$$\lim_{t \to 0} f(t) = 0.$$

In the given relation we replace $t$ by $\frac{t}{4}$, then we do the same in the newly obtained relation and so on, till we get to

$$3f\left(\frac{t}{4}\right) - 4f(t) + f(t) = 3 \cdot \frac{t}{4},$$
$$3f\left(\frac{t}{4^2}\right) - 4f(\frac{t}{4}) + f(t) = 3 \cdot \frac{t}{4^2},$$
$$3f\left(\frac{t}{4^3}\right) - 4f(\frac{t}{4^2}) + f(\frac{t}{4}) = 3 \cdot \frac{t}{4^3},$$
$$\cdots\cdots\cdots\cdots\cdots\cdots\cdots\cdots\cdots\cdots$$
$$3f\left(\frac{t}{4^{n+2}}\right) - 4f(\frac{t}{4^{n+1}}) + f(\frac{t}{4^n}) = 3 \cdot \frac{t}{4^{n+2}}.$$

Summing the last $n + 1$ relations side by side we obtain that

$$3f\left(\frac{t}{4^{n+2}}\right) - f(\frac{t}{4^{n+1}}) - 3f\left(\frac{t}{4}\right) + f(t) = \frac{3t}{4^2}\left(1 + \frac{1}{4} + \ldots + \frac{1}{4^n}\right),$$

for any $n \in \mathbb{N}^*$ and any $t \in \mathbb{R}$. Passing now to the limit and taking into account that $\lim_{y \to 0} f(y) = f(0) = 0$, we deduce that

$$f(t) - 3f\left(\frac{t}{4}\right) = \frac{t}{4}, \ t \in \mathbb{R}.$$

Following now the same steps as we did above, we obtain

$$\begin{aligned}
&f(t) - 3f(\tfrac{t}{4}) = \tfrac{t}{4},\\
&f\left(\frac{t}{4}\right) - 3f(\frac{t}{4^2}) = \frac{t}{4^2}\,|\cdot 3^1\,,\\
&f\left(\frac{t}{4^2}\right) - 3f(\frac{t}{4^3}) = \frac{t}{4^3}\,|\cdot 3^2\,,\\
&\cdots\cdots\cdots\cdots\cdots\cdots\cdots\cdots\cdots\\
&f\left(\frac{t}{4^n}\right) - 3f(\frac{t}{4^{n+1}}) = \frac{t}{4^{n+1}}\,|\cdot 3^n\,.
\end{aligned}$$

We multiply these relations by $3^0, 3^1, 3^2, \ldots, 3^n$, respectively, and sum them, cancelling the like terms to obtain

$$f(t) - 3^{n+1} f\left(\frac{t}{4^{n+1}}\right) = \frac{t}{4}\left[1 + \frac{3}{4} + \ldots + \left(\frac{3}{4}\right)^n\right],$$

i.e.,

$$f(t) - \frac{f(t/4^{n+1})}{t/4^{n+1}} \cdot \left(\frac{3}{4}\right)^{n+1} \cdot t = \frac{t}{4} \cdot \frac{\left[1 - \left(\frac{3}{4}\right)^{n+1}\right]}{1 - \frac{3}{4}}, \; t \neq 0. \tag{6}$$

Passing now to the limit in (6) and taking into account that $\lim\limits_{t \to 0} \dfrac{f(t)}{t} = 1$, we obtain that $f(t) = t$, $t \in \mathbb{R}^*$. As $f(0) = 0$, this shows us that the function we are looking for is $f : \mathbb{R} \to \mathbb{R}$, $f(t) = t, \forall t \in \mathbb{R}$, which is a continuous function. □

We invite readers to study the following

**Problem 15p.** *Find all functions $f : \mathbb{R} \to \mathbb{R}$ that are continuous at zero and satisfy*

$$f(t) - 3f(3t) + 2f(9t) = 10t, \; t \in \mathbb{R}.$$

Now we are going to solve a problem that will suggest another direction for extending the functional equations.

**Problem 15q.** *Find all continuous functions $f : (0, \infty) \to \mathbb{R}$ which satisfy*

$$f(3^x) + f(4^x) = x, \;\; \forall x \in (0, \infty).$$

**Solution.** Let $3^x = y$. This implies $4^x = y^\alpha$, where $\alpha = \log_3 4 > 1$. Thus the given functional equation becomes

$$f(y) + f(y^\alpha) = \log_3 y, \; y \in (0, \infty). \tag{7}$$

In (7) we successively replace $y$ by $y^{\frac{1}{\alpha}}, y^{\frac{1}{\alpha^2}}, \ldots, y^{\frac{1}{\alpha^n}}$ and multiply the newly obtained relations respectively by $(-1)^2, (-1)^3, \ldots, (-1)^{n+1}$. Now we can sum them side by side — this is the basic difference between this problem and Problems 15–15p, where the cancellation of the like terms was direct. Thus we obtain

$$(-1)^{n+1} f\left(y^{\frac{1}{\alpha^n}}\right) + f(y) = \left[\frac{1}{\alpha} - \frac{1}{\alpha^2} + \ldots + (-1)^{n+1}\frac{1}{\alpha^n}\right] \log_3 7, \ \forall n \in \mathbb{N}.$$

Let $n$ be odd. From the previous relation we have that

$$f\left(y^{\frac{1}{\alpha^n}}\right) + f(y) = \frac{\frac{1}{\alpha}\left[1 - \left(-\frac{1}{\alpha}\right)^n\right]}{1 + \frac{1}{\alpha}} \cdot \log_3 y$$

and hence by taking $n \to \infty$ we obtain

$$f(1) + f(y) = \frac{1}{\alpha + 1} \cdot \log_3 y, \ \ y \in (0, \infty).$$

For $y = 1$ we will have $2f(1) = 0$, so $f(1) = 0$ and then

$$f(y) = \frac{1}{\alpha + 1} \log_3 y, \ \ y \in (0, \infty).$$

(The same result can be obtained for $n$ even and $y \neq 1$.)

But $\alpha = \log_3 4$, so the function we were looking for is

$$f(y) = \log_{12} y, \ \ y \in (0, \infty).$$ □

We will end this paragraph with a proposed problem.

**Problem 15r.** *Let $a > 0$, $a \neq 1$ fixed. Find all continuous functions $f : (0, \infty) \to \mathbb{R}$ which satisfy*

$$f(a^x) + f(a^{2x}) = x, \ \ \forall x \in \mathbb{R}.$$

## 15.4 Other Exploration Directions. Functional Inequalities

**Problem 15s.** *Find all continuous functions $f : \mathbb{R} \to \mathbb{R}$ with $f(0) = 0$, $f(1) = 1$ such that*

$$f(x^3) \geq x^2 f(x) \ \text{and} \ \ f(2x) \leq x + f(x), \ \ \forall x \in \mathbb{R}.$$

**Solution.** We exploit both given inequalities in the manner we described in Section 15.2, and then we use the continuity at 0, respectively at 1. Thus $f(x) = x$ is the only solution of the problem. □

The results obtained in Section 15.2 lead us to

**Problem 15t.** *Find all continuous functions $f : (0, \infty) \to \mathbb{R}$ that satisfy*

$$f(2^x) - 2f\left(2^{x+1}\right) + f\left(2^{x+2}\right) \leq 2^{x+1}, \ x \in \mathbb{R} \tag{8}$$

*and*

$$f(3^x) - 2f\left(3^{x+1}\right) + f\left(3^{x+2}\right) \geq 8 \cdot 3^x,\ x \in \mathbb{R}. \tag{9}$$

**Solution.** From the inequality (8) we obtain as in Problem 15h that

$$f\left(\frac{y}{2^{n+2}}\right) - f\left(\frac{y}{2^{n+1}}\right) - f\left(\frac{y}{2}\right) + f(y) \leq \left(1 + \frac{1}{2} + \ldots + \frac{1}{2^n}\right)\frac{y}{2},$$

for any $y \in (0, \infty)$ and any $n \in \mathbb{N}$. As $f$ is continuous at zero, passing to the limit in the previous relation we obtain that

$$f(0) - f(0) - f\left(\frac{y}{2}\right) + f(y) \leq y,$$

i.e.,

$$f(y) - f\left(\frac{y}{2}\right) \leq y,\ \forall\, y \in (0, \infty). \tag{10}$$

This leads us in a similar way to

$$f(y) - f\left(\frac{y}{2^n}\right) \leq y\left(1 + \frac{1}{2} + \ldots + \frac{1}{2^{n-1}}\right),\ y \in (0, \infty),\ n \in \mathbb{N}^*,$$

from which we deduce that

$$f(y) - f(0) \leq 2y \Leftrightarrow f(y) \leq 2y + f(0),\ \forall\, y \in (0, \infty). \tag{11}$$

From the inequality (9) we similarly obtain that

$$f(y) - f\left(\frac{y}{3}\right) \geq \frac{4y}{3},\ \forall\, y \in (0, \infty)$$

and then $f(y) - f(0) \geq 2y$, i.e.,

$$f(y) \geq 2y + f(0),\ \ y \in [0, \infty). \tag{12}$$

Now (11) and (12) lead us to

$$f(y) = 2y + f(0),\ y \in [0, \infty)$$

and so on. □

We now invite readers to compose for themselves other related problems and also to solve the following ones.

**Problem 15u.** *Find all continuous functions* $f : \mathbb{R}_+^* \to \mathbb{R}$, *with* $f(1) = 0$, *that satisfy*

$$f(x^2) \leq f(x) + \ln x \text{ and } f(x^3) \geq f(x) + 2\ln x,\ \forall\, x > 0.$$

**Problem 15v.** *Find all functions* $f : \mathbb{R} \to \mathbb{R}$ *which are continuous at* $x_0 = 0$ *and satisfy*

$$f(2x) - f(x) \leq 3x^2 + x \text{ and } f(3x) - f(x) \geq 8x^2 + 2x.$$

**Problem 15w.** *Find all continuous functions* $f : \mathbb{R} \to \mathbb{R}$, $f(0) = 0$ *that satisfy*

$$f(2x) \geq f(x) \text{ and } f(3x) \leq 2x + f(x),\ \forall\, x \in \mathbb{R}.$$

## 15.5 Bibliographical Comments

Problem 15 was published in *Gazeta Matematica* no. 1/1988 as problem 21333*, having M. Cavachi as author. No solution of this problem was published in that journal.

Problem 15c was published in *Gazeta Matematica* no. 1/1989 as problem 21671* and having V. Berinde as author. Its solution, which is entirely reproduced here, was published in no. 1/1990 of the same journal.

Problem 15j was published in *Gazeta Matematica* no. 9/1994 as problem 23073.

Other variants of this class of problems were proposed by the author at many editions of the county round, Maramures, of the National Mathematics Olympiad and at the Intercounty Mathematics Contest "Gr.C. Moisil".

Problem 15k appeared in *Gazeta Matematica* no. 9/1987 as problem 21221* having M. Chirita as author, while Problem 15l is the second problem proposed at the county round of the National Mathematics Olympiad, Bucharest, $5^{th}$ of March 1994, and was published in *Gazeta Matematica* no. 4/1994, p. 152.

Problem 15s was proposed by F. Rotaru and published in *Gazeta Matematica* no. 4/1994 as problem 22985. Problem 15w belongs to the same author and was published as problem 22872 in *Gazeta Matematica* no. 9/1993.

Problem 15q was proposed by M. Ionescu and appeared in *Gazeta Matematica* no. 8/1994 as problem C:1571, in the column "Problems for the Annual Contest of the Solvers".

Problem 15r was proposed by M. Chirita and V. Matrosenco at the county round of the National Mathematics Olympiad, Bucharest, 1988, and then was published in *Gazeta Matematica* no. 4/1998.

Problem 15q was proposed by M. Chirita at the National Mathematics Camp in 1998 and it was then published in *Gazeta Matematica* no. 10/1998, pp. 387, while Problem 15r appeared as problem 23709 in *Gazeta Matematica* no. 3/1997, proposed by M. Ursarescu.

## 15.6 References

1. **V. Arsinte,** *Elementary Problems in Integral Calculus.* (Romanian), Editura Universitatii Bucuresti, 1995

2. **V. Berinde,** *How to Stimulate Engineering Invention by Means of Mathematical Approaches.* Proceed. of Internat. Conf. "Teaching Mathematics for Industry" (J. Cerny, ed.), Czech Technical University, Prague, 1996, pp. 19–30

3. **St. Poka, Maria-Adriana Poka,** *On a Problem from Gazeta Matematica.* (Romanian), Gazeta Matematica, **94** (1989), no. 2, 54–56

# Theme 16

# An Extension of the Leibniz–Newton Formula

In this section we are going to present a slight extension of the Leibniz–Newton formula for the computation of the definite integral, an extension that can be applied also in the case of some integrable functions which do not admit primitives.

## 16.1 Source Problem

The investigations in this section will have as a starting point

**Problem 16.** *Let* $f : \mathbb{R} \to \mathbb{R},\ f(x) = \dfrac{1}{3+\cos x},\ x \in \mathbb{R}.$

*a) Find all primitives of* $f$ *on* $[0, \pi)$;

*b) Find all primitives of* $f$ *on* $[0, 2\pi]$ *and calculate* $\int\limits_0^{2\pi} f(x)\,dx.$

**Solution.** a) As $\cos x \in [-1, 1]$, $\forall x \in \mathbb{R}$, we have $3 + \cos x > 0$, $\forall x \in \mathbb{R}$, and so $f$ is continuous on $\mathbb{R}$. This implies that $f$ admits primitives on $\mathbb{R}$ (and on any interval of the real axis).

By changing the variable $\tan\dfrac{x}{2} = t$, which stands for any interval $I \subset \mathbb{R}$ that does not contain points like $(2k+1)\pi,\ k \in \mathbb{Z}$, we obtain that

$$\int \frac{dx}{3+\cos x} = \frac{1}{\sqrt{2}} \arctan \frac{\tan\frac{x}{2}}{\sqrt{2}} + c,\quad x \in I.$$

This shows that

$$F_1(x) = \frac{1}{\sqrt{2}} \arctan \frac{\tan\frac{x}{2}}{\sqrt{2}} + c \tag{1}$$

is a primitive of the function $f$ on $[0, \pi)$ and also on the interval $(\pi, 2\pi]$.

b) As $F_1$ is not defined at $x = \pi$, $F_1$ cannot be a primitive of $f$ on $[0, 2\pi]$. Such a primitive, say $F(x)$, will look like

$$F(x) = \begin{cases} \dfrac{1}{\sqrt{2}} \arctan \dfrac{\tan \frac{x}{2}}{\sqrt{2}} + c_1\,, & x \in [0, \pi), \\ c_2\,, & x = \pi, \\ \dfrac{1}{\sqrt{2}} \arctan \dfrac{\tan \frac{x}{2}}{\sqrt{2}} + c_3\,, & x \in (\pi, 2\pi]. \end{cases} \tag{2}$$

We determine the constants $c_1, c_2$ and $c_3$ so that $F$, given by (2), is a primitive of $f$ on $[0, 2\pi]$. Any primitive is differentiable and therefore continuous. Considering the continuity conditions of $F$ at $\pi$, we obtain

$$F(\pi - 0) = F(\pi) = F(\pi + 0), \tag{3}$$

where

$$F(\pi - 0) = \lim_{\substack{x \to \pi \\ x < \pi}} F(x) \text{ and } F(\pi + 0) = \lim_{\substack{x \to \pi \\ x > \pi}} F(x).$$

As $x < \pi \Rightarrow \dfrac{x}{2} < \dfrac{\pi}{2} \Rightarrow \tan \dfrac{x}{2} > 0$ and respectively $x > \pi \Rightarrow \dfrac{x}{2} > \dfrac{\pi}{2} \Rightarrow \tan \dfrac{x}{2} < 0$, we deduce that

$$\lim_{\substack{x \to \pi \\ x < \pi}} \tan \frac{x}{2} = +\infty \text{ and } \lim_{\substack{x \to \pi \\ x > \pi}} \tan \frac{x}{2} = -\infty$$

and therefore

$$\lim_{\substack{x \to \pi \\ x < \pi}} \arctan \frac{\tan \frac{x}{2}}{\sqrt{2}} = \frac{\pi}{2}, \text{ while } \lim_{\substack{x \to \pi \\ x > \pi}} \arctan \frac{\tan \frac{x}{2}}{\sqrt{2}} = -\frac{\pi}{2},$$

which leads us to

$$F(\pi - 0) = \frac{\pi}{2\sqrt{2}} + c_1 \text{ and } F(\pi + 0) = -\frac{\pi}{2\sqrt{2}} + c_2.$$

Denoting $c = c_2$, by (3) we obtain $c_1 = -\dfrac{\pi}{2\sqrt{2}} + c$ and $c_3 = \dfrac{\pi}{2\sqrt{2}} + c$. Replacing $c_1$, $c_2$ and $c_3$ in (2) we have that

$$F(x) = \begin{cases} \dfrac{1}{\sqrt{2}} \arctan \dfrac{\tan \frac{x}{2}}{\sqrt{2}} - \dfrac{\pi}{2\sqrt{2}} + c\,, & x \in [0, \pi), \\ c\,, & x = \pi, \\ \dfrac{1}{\sqrt{2}} \arctan \dfrac{\tan \frac{x}{2}}{\sqrt{2}} + \dfrac{\pi}{2\sqrt{2}} + c\,, & x \in (\pi, 2\pi], \end{cases} \tag{2'}$$

easily proven to be a primitive of $f$ on $[0, 2\pi]$.

In order to calculate $\int\limits_0^{2\pi} f(x)dx$ we use now the Leibniz–Newton formula (see for example [7]), with the following general statement.

**Theorem 1.** *Let $f : [a, b] \to \mathbb{R}$ be such that:*

*(i) $f$ is integrable on $[a, b]$;*

*(ii) $f$ admits primitives on $[a, b]$;*

*(iii) $F$ is an arbitrary primitive of $f$ on $[a, b]$.*

*Then*

$$\int\limits_a^b f(x)dx = F(b) - F(a).$$

In the case of our problem we will obtain

$$\int\limits_0^{2\pi} \frac{dx}{3+\cos x} = F(2\pi) - F(0) = \frac{\pi}{2\sqrt{2}} + c - \left(-\frac{\pi}{2\sqrt{2}} + c\right) = \frac{\pi}{\sqrt{2}},$$

since $\arctan \frac{\tan 0}{\sqrt{2}} = 0$. □

We invite readers to solve by themselves the following problems, which have the same statement as Problem 16.

**16a.** $f(x) = \dfrac{1}{3 + \sin x + \cos x}$; **16b.** $f(x) = \dfrac{1}{2 + \sin x}$; **16c.** $f(x) = \dfrac{1}{2 + \cos x}$.

## 16.2 A Close Examination of the Problem and its Solution. The Generalized Leibniz–Newton Formula

If we had used the "primitive" $F$ given by (1) when calculating the definite integral in Problem 16, we would have obtained

$$\int\limits_0^{2\pi} \frac{dx}{3+\cos x} = F_1(2\pi) - F_1(0) = 0 - 0 = 0.$$

As $\dfrac{1}{3+\cos x} > 0,\ \forall x \in [0, 2\pi]$, this result would have contradicted the fact that $\int\limits_0^{2\pi} \dfrac{dx}{3+\cos x} > 0$.

This shows that, in order to apply the Leibniz–Newton formula, all the (i)–(iii) assumptions of the theorem have to be satisfied. Using $F_1$ would have meant exactly ignoring (iii).

A question arises related to this problem (a question that not everybody can answer firmly): why is $F_1$, although a "legitimate" primitive of $f$ on $[0, \pi)$, as we have proved in 16.1, however not a primitive on $\mathbb{R}$?

The answer is to be found in the very method we used to calculate the primitives of $f$. This implies the change of variables $tan\frac{x}{2} = t$ which, as we have showed, stands only for intervals free from points like $(2k+1)\pi,\ k \in \mathbb{Z}$. Therefore, although $f$ has primitives on $\mathbb{R}$, we cannot obtain **directly** a primitive on $\mathbb{R}$ using the method described above (and the only one available, actually).

As the following problem also shows, we can have similar situations even if we use integration by parts instead of variable change.

**Problem 16d.** *Find a primitive of* $f : \mathbb{R} \to \mathbb{R}$,

$$f(x) = \ln\left(x + \sqrt{x^2+4}\right),\ x \in \mathbb{R},$$

*a) on the interval* $(2,\infty)$; *b) on the interval* $[2,\infty)$.

**Solution.** We integrate by parts, denoting $u = \ln\left(x+\sqrt{x^2-4}\right)$, $v' = 1$, $x \in (2,\infty)$, and thus we obtain

$$u' = \frac{1}{\sqrt{x^2-4}},\quad v = x,$$

so

$$F_1(x) = x\ln\left(x+\sqrt{x^2-4}\right) - \sqrt{x^2-4}+c$$

is a primitive of the given function on $(2,\infty)$. □

Although $F_1$ **is defined on** $[\mathbf{2},\infty)$, we cannot state that $F_1$ is a primitive of $f$ on $(2,\infty)$. That is because the integration by parts requires a continuous derivative and hence cannot be applied on the entire interval $[2,\infty)$. Therefore, we are going to search for a primitive on $[2,\infty)$, of the form

$$F(x) = \begin{cases} F_1(x)+c_1\,, & x \in (2,\infty),\\ c_2\,, & x = 2.\end{cases}$$

Although here we have in fact $c_1 = c_2$ , that is, $F(x) = F_1(x)+c,\ x \in [2,\infty)$ is a primitive of $f$ on $[2,\infty)$, to deduce that just starting from the fact that $F_1$ was a primitive of $f$ on $(2,\infty)$, without the above approach, would have been incorrect. And that is because a primitive has to be defined not on a union of intervals but on an interval.

Thus the following question arises: *if $F$ is a "primitive" on a union of intervals, then can it be used directly in a Leibniz-Newton-like formula* ? A positive answer is given by

**Theorem 2.** *Let $f : [a,b] \to \mathbb{R}$ be such that:*

*(i) $f$ is integrable on $[a,b]$;*

*(ii) $f$ admits primitives on $[a,b]$.*

*Let $c \in (a,b)$ and $F : [a,b]\backslash\{c\} \to \mathbb{R}$ be a differentiable function such that*

$$F'(x) = f(x),\quad \forall\, x \in [a,b]\backslash\{c\}. \tag{4}$$

*Then $F$ has finite lateral limits at $c$ and*

$$\int_a^b f(x)\,dx = F(b) - F(a) + F(c-0) - F(c+0).$$

**Proof.** Let $G$ be a primitive of $f$ on $[a,b]$. As $F'(x) = G'(x) = f(x)$, $\forall\, x \in [a,c)$, there is a constant $c_1$ such that $F(x) = G(x) + c_1$, $\forall\, x \in [a,c)$.

In a similar way we obtain that there is a constant $c_2$ (generally not equal to $c_1$) such that $F(x) = G(x) + c_2$, $\forall\, x \in (c,b]$.

As $G$ is a primitive on $[a,b]$, $G$ is also continuous on $[a,b]$ and so $G$ has a limit at $x = c$, i.e., $G$ has lateral limits at $c$:

$$F(c-0) = G(c) + c_1 \text{ and } F(c+0) = G(c) + c_2.$$

Now applying the Leibniz–Newton formula for the primitive $G$ we obtain

$$\int_a^b f(x)\,dx = G(b) - G(a) = F(b) - F(a) + c_1 - c_2$$

and, since

$$c_1 - c_2 = F(c-0) - F(c+0),$$

we get exactly to the conclusion of the theorem. □

**Remarks.** 1) Theorem 2 is an extension of the Leibniz–Newton formula. If $F$ is continuous at $c$, then $F(c-0) = F(c+0)$ and Theorem 2 leads us immediately to Theorem 1;

2) The generalized Leibniz–Newton formula can now be used in solving Problem 16.

Indeed, the function $F_1$ given by (1) has the property

$$F_1'(x) = \frac{1}{3+\cos x}, \quad \forall\, x \in [0,2\pi]\backslash\{\pi\}$$

and then

$$\begin{aligned}\int_0^{2\pi} \frac{dx}{3+\cos x} &= F_1(2\pi) - F_1(0) + F_1(\pi-0) - F_1(\pi+0)\\ &= 0 - 0 + \frac{\pi}{2\sqrt{2}} - \left(-\frac{\pi}{2\sqrt{2}}\right) = \frac{\pi}{\sqrt{2}}.\end{aligned}$$

3) Only the situation when $c$ is an interior point of the $[a,b]$ interval is considered in Theorem 2. A more general result can be formulated (see [2]–[6], [8], [9]). In this case there is a finite (or infinite but of null measure) set of points $c \in [a,b]$

on which $F$ is not defined. Starting from Theorem 2, in [2]–[6] we have introduced the concept of *a.e. primitive* as well as the one of *a.e. integral*, which extend the usual notions of primitive and integral. We will not present them here, as we do not want to exceed the elementary framework we have respected till now, still we will mention that by means of these notions one has the possibility to gather in the same class the Riemann integrable functions which do not admit primitives and the functions which admit primitives but are not Riemann integrable. All these are integrable in the sense **almost everywhere** (see [6]).

4) Assumption (ii) in Theorem 2 is not necessary in order to obtain the conclusion. Instead of that, it suffices that there exists $F : [a,b] \to \mathbb{R}$ such that (4) holds.

**Corollary.** *Let $f : [a,b] \to \mathbb{R}$ be such that $f$ is integrable on $[a,b]$ and there exist $c \in (a,b)$ and a function $F : [a,b]\backslash\{c\} \to \mathbb{R}$ such that it is differentiable on* $[a,b]\backslash\{c\}$ and

$$F'(x) = f(x), \quad \forall\, x \in [a,b]\backslash\{c\}.$$

*Then $F$ has finite lateral limits at $c$ and*

$$\int_a^b f(x)\,dx = F(b) - F(a) + F(c-0) - F(c+0). \qquad \square$$

**Problem 16e.** *Compute $\int_0^2 f(x)\,dx$, where*

$$f : [0,2] \to R, \quad f(x) = \begin{cases} 1, & x \in [0,2] \setminus \{1\}, \\ -1, & x = 1. \end{cases}$$

**Solution.** The given function $f$ is integrable but does not admit primitives, since it does not have Darboux' property. So we cannot use the classical Leibniz–Newton formula here. But we can apply the corollary, since the function $F(x) = x$ has the property that $F'(x) = f(x)$, $\forall\, x \in [0,2] \setminus \{1\}$. Therefore

$$\int_0^2 f(x)\,dx = F(2) - F(0) + F(1-0) - F(1+0) = 2 - 0 + 1 - 1 = 2.$$

## 16.3 Conclusions

Although the gains brought by the investigations in this section are not outstanding — this meaning that we have not obtained so many new exquisite problems — we still have to remark that the light our research has put on the primitives and integral calculus for some special functions. The major benefit is the (re)discovery of the generalized Leibniz–Newton formula and of other more general results that are mentioned in our cited papers.

The creativity involved in this section has a strong corrective role, calling one's attention to some mistakes that frequently appear in solving problems of this type. No wonder then that, attending the $9^{th}$ edition of the International Congress on Mathematical Education, Tokyo, 31 July - 6 August 2000, I could see a presentation of a European mathematician where the author expressed his puzzle that $F_1(x)$ given by (1), although the primitive of such a "quiet" function as $f(x) = \dfrac{1}{3+\cos x}$, has the "bizarre" behaviour one can notice when plotting its graph by means of a computer. I certainly would have enlightened him if I had had a copy of this paper with me there.

## 16.4 Bibliographical Comments

Problem 16 was part of the second test (18 July) of the entrance exam at technical universities in 1988. It was then published in *Gazeta Matematica* no. 11–12/1988, p. 452. It was preferred as the source problem because of the simpler computations needed to solve it. A similar problem, 16a, had been proposed much earlier at the final round of the National Mathematics Olympiad in 1983, its author being I. Branza. In that variant, see [10],

$$f(x) = 3/(3 + \sin x + \cos x).$$

Problem 16b is taken from [1].

Problem 16d was published in *Gazeta Matematica* no. 10–11–12/1990, as problem 22223, its author being V. Berinde. Similar problems have been proposed by the same author at the Maramures county round of the National Mathematics Olympiad in 1988 and 1990, see [10].

Theorem 2 is a reformulation of the problem 22126* from *Gazeta Matematica* no. 6–7/1990, author V. Berinde. The same problem was proposed at the Maramures county round of the National Mathematics Olympiad, in the years 1990–1993.

All the results of this kind were first studied in detail by Ch. Hermite in the 1850s.

## 16.5 References

1. **L. Arama, T. Morozan,** *Problems Book of Differential and Integral Calculus.* (Romanian), Editura Tehnica, 1967

2. **V. Berinde,** *Primitivability and Almost Everywhere Integrability.* (Romanian), The County Symposium of Scientifical Papers "Creativity and Efficiency in Education", $4^{th}$ Edition, Baia Mare, 28 March 1989

3. **V. Berinde,** *Primitivability and Almost Everywhere Integrability.* (Romanian), The Methodical-Scientific Session "Creativity and Efficiency in Education", The National Round, Iasi, 2–3 June 1989

4. **V. Berinde,** *A Notion of Almost Everywhere Primitive.* Bul. Stiint. Univ. Baia Mare Ser. B, **7** (1991), 35–42

5. **V. Berinde,** *Remarks on the Leibniz-Newton Formula.* (Romanian), in MATEMATICA, Synthesia and Tests, Univ. of Baia Mare, 1991, pp. 13–16

6. **V. Berinde,** *A generalization of the Riemann Integral.* Bul. Stiint. Univ. Baia Mare Ser. B, **8** (1992), 1–10

7. **N. Boboc, I. Colojoara,** *Elements of Analysis. Textbook for the* $12^{th}$ grade. (Romanian), Editura Didactica si Pedagogica, 1987

8. **N. Bourbaki,** *Fonctions d'une variable reéle (Théorie élémentaire).* Herman, 1958

9. **Gh. Siretchi,** *Differential and Integral Calculus.* (Romanian), Editura Stiintifica si Enciclopedica, vol. 1–2, 1985

10. **I. Tomescu,** *Problems Given at the Mathematics Olympiad for High Schools.* (Romanian), Editura Stiintifica, 1992

11. **S. Lang,** *A First Course in Calculus*, 5th ed., Springer-Verlag, New York, 1987

# Theme 17

# A Measurement Problem

In some books on recreational mathematics, but also in mathematics journals and school competitions, one can sometimes meet problems asking to find a certain quantity (mass, volume), without using a standard measurement unit. A problem of this kind is the source problem of this section.

## 17.1 Source Problem. Some Exploratory Directions

**Problem 17.** *A* 12 *liter pot is filled with wine. Remove* 4 *liters of this quantity to a* 7 *liter pot, using a* 5 *liter pot.*

**Solution.** We indicate the operations done step by step, using the notation "$v_1 \to v_2$" which says that we pour from the pot $v_1$ into the pot $v_2$. The result of the operation is written on the following line of the matrix.

| Step | Pot 1 (12l) | Pot 2 (7l) | Pot 3 (5 l) | Operation |
|---|---|---|---|---|
| I | 12 | 0 | 0 | 1→ 2 |
| II | 5 | 7 | 0 | 2→ 3 |
| III | 5 | 2 | 5 | 3→ 1 |
| IV | 10 | 2 | 0 | 2→ 3 |
| V | 10 | 0 | 2 | 1→ 2 |
| VI | 3 | 7 | 2 | 2→ 3 |
| VII | 3 | 4 | 5 | 3→ 1 |
| VIII | 8 | 4 | 0 | |

□

**Problem 17a.** *Divide* 8 *liters into* 2 *equal parts using a* 5 *liter pot and a* 3 *liter pot.*

**Solution.** We follow the method used in solving Problem 17, writing the operations and their results in the next matrix.

| Step | Pot 1 (8 l) | Pot 2 ( 5 l) | Pot 3 (3 l) | Operation |
|---|---|---|---|---|
| I | 8 | 0 | 0 | 1→ 2 |
| II | 3 | 5 | 0 | 2→ 3 |
| III | 3 | 2 | 3 | 3→ 1 |
| IV | 6 | 2 | 0 | 2→ 3 |
| V | 6 | 0 | 2 | 1→ 2 |
| VI | 1 | 5 | 2 | 2→ 3 |
| VII | 1 | 4 | 3 | 3→ 1 |
| VIII | **4** | **4** | 0 | |

□

**Remark.** Comparing the data in the previous two problems, the order and the number of operations, we realize that they work basically on the same principle. But what is this principle? The answer is given by the next problem, which is also a small "factory" of problems of the same kind as Problems 17 and 17a.

**Problem 17b.** *A pot of* $m+n$ *liters is full. Remove* $2(m-n)$ *liters in an* $m$ *liter pot, using an* $n$ *liter pot* $(m, n \in \mathbb{N}^*,\ n < m \leq 2n)$.

**Solution.**

| Step | Pot 1 $(m+n)$ l | Pot 2 $m$ l | Pot 3 $(n)$ l | Operation |
|---|---|---|---|---|
| I | $m+n$ | 0 | 0 | 1→ 2 |
| II | $n$ | $m$ | 0 | 2→ 3 |
| III | $n$ | $m-n$ | $n$ | 3→ 1 |
| IV | $2n$ | $m-n$ | 0 | 2→ 3 |
| V | $2n$ | 0 | $m-n$ | 1→ 2 |
| VI | $2n-m$ | $m$ | $m-n$ | 2→ 3 |
| VII | $2n-m$ | $2(m-n)$ | $n$ | 3→ 1 |
| VIII | $3n-m$ | $\mathbf{2(m-n)}$ | 0 | |

Thus in the second pot we got the $2(m-n)$ liters required by the problem. □

**Comments**. Step III. As $m > n$, we have $m-n > 0$.
Step V. Since $m \leq 2n$, it results that $2n-m \geq 0$, and so we can pour from 1 into 2.
Step VI. The pot 3 will be filled if we pour $n-(m-n) = 2n-m$ liters from pot 2. In pot 2 are left exactly $m-(2n-m) = 2(m-n)$ liters.

**Remarks.** 1) By taking $n = 7,\ m = 5$ in Problem 17b, we get exactly Problem 17 and by taking $n = 5,\ m = 3$ we get Problem 17a.

2) As an immediate generalization of Problem 17 and 17a, Problem 17b is not very important by itself. What is important, though, is that it provides as many problems as we wish, of this certain type. The teachers who want to make sure that their students will have a good approaching technique for this class of

problems have just to take 2 numbers $m$ and $n$, verifying

$$n < m \text{ and } m < 2n,$$

and they will instantly get a new problem.

3) We suggest that reader complicate the problem by adding two auxiliary pots to the starting one and the one where the wanted quantity has to be poured. We also recommend that they approach problems of mass measurement, without standard units.

4) By a more careful analysis of the measurement operations (of the 4 liters in Problem 17), we notice that it is not necessary that the source pot (the bigger one) is completely full, as the greatest quantity poured from it into the other two was 9 liters and this happened during steps VI and VII. Therefore we can solve the same problem with a smaller initial quantity, but not less than 9 liters. Thus we get the following problem.

**Problem 17c.** *From a pot containing* 9 *liters of water, remove* 4 *liters into a* 5 *liter pot, using a* 7 *liter pot.*

**Solution.** The operations are shown in the matrix

| Step | Pot 1 ($\geq 9$ l) | Pot 2 (7 l) | Pot 3 (5 l) | Operation |
|---|---|---|---|---|
| I | 9 | 0 | 0 | $1 \to 2$ |
| II | 2 | 7 | 0 | $2 \to 3$ |
| III | 2 | 2 | 5 | $3 \to 1$ |
| IV | 7 | 2 | 0 | $2 \to 3$ |
| V | 7 | 0 | 2 | $1 \to 2$ |
| VI | 0 | 7 | 2 | $2 \to 3$ |
| VII | 0 | 4 | 5 | $3 \to 1$ |
| VIII | 5 | 4 | 0 | $2 \to 3$ |
| IX | 5 | 0 | **4** | |

□

The following problems are solved in the same manner as the previous ones.

**Problem 17d.** *From a pot containing* 10,5 *liters of water, pour* 4 *liters into a* 5 *liter pot, using a* 7 *liter pot.*

**Problem 17e.** *From a pot containing* 8,5 *liters, pour* 4 *liters into a* 5 *liter pot, using a* 3 *liter pot.*

We can formulate now a problem of the same type as 17b, but less restrictive than that.

**Problem 17f.** *A pot has more than* $2m-n$ *liters of wine. Pour* $2(m-n)$ *liters into an* $m$ *liter pot, using an* $n$ *liter pot* ($m, n \in \mathbb{N}^*$, $n < m \leq 2n$).

**Solution.** Let $x > 2m - n$ liters be the quantity in pot 1. We have

| Step | Pot 1 ($> 2m - n$ l) | Pot 2 ($m$ l) | Pot 3 ($n$ l) | Operation |
|---|---|---|---|---|
| I | $x$ | 0 | 0 | 1→ 2 |
| II | $x - m$ | $m$ | 0 | 2→ 3 |
| III | $x - m$ | $m - n$ | $n$ | 3→ 1 |
| IV | $x + n - m$ | $m - n$ | 0 | 2→ 3 |
| V | $x + n - m$ | 0 | $m - n$ | 1→ 2 |
| VI | $x + n - 2m$ | $m$ | $m - n$ | 2→ 3 |
| VII | $x + n - 2m$ | $2(m - n)$ | $n$ | 3→ 1 |
| VIII | $x + 2n - 2m$ | **2(m-n)** | 0 | |

At step VI we pour $n - (m - n)$ liters from pot 2 into pot 3, so in pot 2 there remain

$$m - (2n - m) = 2(m - n) \text{ liters.}$$ □

**Comments.** Step I. Since $x - m > 2m - n - m = m - n > 0$, we can pour $m$ liters from 1 into 2.
Step V. As $x - m > m - n$, i.e., $x > 2m - n$, we have $x + m - n$ liters in pot 1 and so we can pour $m$ liters into pot 2.

## 17.2 Bibliographical Comments

Problem 17 was proposed by I. Lintes in *The Supplement of Gazeta Matematica* as no. E:420 and was republished in *Gazeta Matematica* no. 9/1980, p. 393 (we replaced "water" by "wine").

Problem 17a is problem 53, "At a concrete station" (Romanian) from [2], whose text we simplified here.

## 17.3 References


1. **E. Guran,** *Recreative Mathematics.* (Romanian), Editura Junimea, 1985
2. **E. Guran,** *Problems and Exercices of Recreative Mathematics.* (Romanian), Editura Lumina, 1991

# Theme 18

# A Class of Discontinuous Functions Admitting Primitives

The topic of this section, i.e., the existence of primitives for discontinuous functions, is well represented in the Romanian elementary mathematics literature. After the textbook of analysis for the 12th form [5], containing a generous approach of the topic of primitives, was published, a real avalanche of exercises and problems regarding the existence of primitives for discontinuous functions invaded elementary mathematics journals as well as mathematics competitions. In addition, several remarks, notes, survey articles or conference presentations were dedicated to the same topic. Therefore it would be quite impossible to have here a complete survey of the material. This is the reason why we limit ourselves to a very narrow class of such problems, already approached in our paper [4].

## 18.1 Source Problem

**Problem 18.** *Using the fact that any continuous function on an interval $I$ has a primitive, show that the function*

$$f(x) = \begin{cases} \cos\dfrac{1}{x}, & \textit{if } x \in \mathbb{R}\backslash\{0\}, \\ 0, & \textit{if } x = 0, \end{cases}$$

*has primitives on $\mathbb{R}$.*

**Solution.** Let $I \subset \mathbb{R}$ be an interval that does not contain 0. Then on this interval we have

$$\int \cos\frac{1}{x}dx = -\int x^2\left(\sin\frac{1}{x}\right)' dx$$

and hence integrating by parts we get

$$\int \cos \frac{1}{x} dx = -x^2 \sin \frac{1}{x} + 2 \int x \sin \frac{1}{x} dx. \tag{1}$$

Let $g : \mathbb{R} \to \mathbb{R}$ be the function

$$g(x) = \begin{cases} x \sin \dfrac{1}{x}, & \text{for } x \in \mathbb{R}\backslash\{0\}, \\ 0, & \text{for } x = 0. \end{cases}$$

Then, considering the double inequality

$$-|x| \leq x \sin \frac{1}{x} \leq |x|, \; x \in \mathbb{R}\backslash\{0\},$$

we deduce that $g$ is continuous at zero, and hence $g$ is continuous on $\mathbb{R}$. Hence $g$ admits primitives on $\mathbb{R}$. Let $G : \mathbb{R} \to \mathbb{R}$ be one of its primitives. We have

$$G'(x) = x \sin \frac{1}{x}, \; \forall x \in \mathbb{R}^*$$

and by (1) we deduce that any primitive of $f$ will have the form

$$F(x) = \begin{cases} -x^2 \sin \dfrac{1}{x} + 2G(x) + c_1, & \text{for } x \in \mathbb{R}\backslash\{0\} \\ c_2, \text{for } x = 0. \end{cases} \tag{2}$$

In order to be a primitive, $F$ given by (2) must be differentiable and hence continuous. In particular, $F$ must be continuous at zero, which gives

$$\lim_{x \to 0} F(x) = F(0),$$

i.e.,

$$-\lim_{x \to 0} x^2 \sin \frac{1}{x} + 2 \lim_{x \to 0} G(x) + c_1 = c_2.$$

But $G$ is continuous (being a primitive) and hence

$$\lim_{x \to 0} G(x) = G(0).$$

On the other hand, using the double inequality

$$-\left|x^2\right| \leq x^2 \sin \frac{1}{x} \leq \left|x^2\right|, \; x \in \mathbb{R}^*$$

it results that $\lim\limits_{x \to 0} x^2 \sin \dfrac{1}{x} = 0$. Therefore

$$c_1 + 2G(0) = c_2$$

and denoting $c_1 = c$ we deduce that any primitive of $f$ has the form

$$F(x) = \begin{cases} -x^2 \sin \frac{1}{x} + 2G(x) + c_1, & \text{for } x \in \mathbb{R}\backslash\{0\}, \\ c_2, & \text{for } x = 0. \end{cases} \tag{2'}$$

We shall prove that $F : \mathbb{R} \to \mathbb{R}$ given by (2′) is a primitive of $f$ on $\mathbb{R}$. Indeed, for any $x \in \mathbb{R}\backslash\{0\}$, $F$ is differentiable and

$$F'(x) = f(x).$$

We study now the differentiability of $F$ at zero. According to the definition, we have to compute

$$\lim_{x\to 0} \frac{F(x) - F(0)}{x - 0} = \lim_{x\to 0} \frac{-x^2 \sin \frac{1}{x} + 2G(x) + c - 2G(0) - c}{x}$$

which is equal to

$$-\lim_{x\to 0} x \sin \frac{1}{x} + 2 \lim_{x\to 0} \frac{G(x) - G(0)}{x - 0}.$$

But $G$ is a primitive of $g$ on $\mathbb{R}$, hence $G$ is differentiable at zero and $G'(0) = g(0) = 0$. This shows that

$$\lim_{x\to 0} \frac{G(x) + G(0)}{x - 0} = G'(0) = g(0) = 0$$

and hence

$$F'(0) = 0 = f(0)$$

which proves that $F$ is differentiable at zero and, for any $x \in \mathbb{R}$,

$$F'(x) = f(x).$$

Therefore $F$ is a primitive of $f$ on $\mathbb{R}$. □

We can similarly solve

**Problem 18a.** *Show that the function* $f : \mathbb{R} \to \mathbb{R}$

$$F(x) = \begin{cases} \sin \frac{1}{x}, & \textit{for } x \in \mathbb{R}\backslash\{0\}, \\ 0, & \textit{for } x = 0, \end{cases}$$

*admits primitives on* $\mathbb{R}$.

**Remarks.** 1) If we consider the sequences $(x_n'), (x_n'')$, given by

$$x_n' = \frac{1}{2n\pi}, \quad n \in \mathbb{N}^* \quad \text{and} \quad x_n'' = \frac{2}{(2n+1)\pi}, \quad n \in \mathbb{N}^*,$$

respectively, then $\lim_{n\to\infty} x'_n = \lim_{n\to\infty} x''_n = 0$ and

$$\lim_{n\to\infty} \cos\frac{1}{x'_n} = \lim_{n\to\infty} \cos 2n\pi = 1 \neq 0 = \lim_{n\to\infty} \cos(2n+1)\pi = \lim_{n\to\infty} \cos\frac{1}{x''_n}.$$

This shows that the function $f$ in Problem 18 has no limit at zero and hence it is not continuous at zero. Therefore, the functions $f$ in Problems 18 and 18a are examples of discontinuous functions admitting primitives. It is well known that any continuous function also admits primitives. Considering the results obtained above, the reverse is not generally true: if $f$ has primitives it does not result that $f$ is continuous;

2) If we analyze the way of obtaining (1) integrating by parts, we can notice that the same method could be applied in the case when $f$ has a more complicated expression. In the next paragraph we will be interested in exploring how the expression of $f$ could be extended in order to preserve its main properties, i.e., the fact that $f$ is a discontinuous function admitting a primitive.

## 18.2 Exploring the Class of Discontinuous Functions Admitting Primitives

If we carefully analyze Problem 18 and its solution we notice many interesting aspects. First, the function $\sin\frac{1}{x}$ is bounded on $\mathbb{R}^*$. This property was essential to deduce that

$$\lim_{n\to\infty} x^k \sin\frac{1}{x} = 0 \quad (k > 0).$$

Also, the above function is periodic, is not defined at zero and does not have limit at zero. The great majority of the papers dedicated to this topic ([1], [10], [11], [12]) exploit mainly the periodicity and boundedness of $f$. In the following we shall ignore the periodicity, even if in this way we lose a large number of exploratory directions. Our main aim here will be to find a class as large as possible of discontinuous functions admitting primitives. Due to the fact that it is a good algorithm and has a high level of generality, the solution of Problem 18 will provide readers with a clear and consistent method for solving similar problems.

To this end, we consider an interval $I \subset \mathbb{R}$ and let $J$ denote one of the intervals $(-\infty, \infty)$, $(-\infty, 0]$ or $[0, \infty)$.

Let $p, q$ and $r$ be three functions, $p : J \to \mathbb{R}$, $q$, $r : I \to \mathbb{R}$ such that $q(I) \subset J$.

Exceptionally, in this section we shall not indicate all the steps of our successive explorations and we shall limit ourselves to presenting just the final results, in order to prevent the size of this section from becoming too large. The most general result of this kind is contained in the next theorem.

**Theorem 1.** *Let the functions $p, q$ and $r$ satisfy:*

*a) there exists a unique point $x_0 \in I$ with $q(x_0) = 0$;*

*b) $q$ and $r$ are differentiable on $I \backslash \{x_0\}$ and have continuous derivatives on $I \backslash \{x_0\}$;*

*c) $p$ is bounded and differentiable on $J$;*

*d) $\lim\limits_{x \to x_0} \dfrac{r(x) \cdot q^2(x)}{x - x_0} = 0$;*

*e) $\lim\limits_{x \to x_0} [r(x) \cdot q^2(x)]' = 0$.*

*Then the function $f : I \to \mathbb{R}$ given by*

$$f(x) = \begin{cases} r(x) \cdot q'(x) \cdot p'\left(\dfrac{1}{q(x)}\right), & \text{for } x \neq x_0, \\ 0, & \text{for } x = x_0, \end{cases} \tag{3}$$

*has primitives on $I$.*

**Proof.** On any interval that does not contain $x_0$ we have

$$\int r(x)q'(x)p'\left(\frac{1}{q(x)}\right) dx = -\int \left[r(x) \cdot q^2(x)\right] \cdot \left[p\left(\frac{1}{q(x)}\right)\right]' dx$$

and integrating by parts we obtain

$$\int r(x)q'(x)p'\left(\frac{1}{q(x)}\right) dx = -r(x)q^2(x)p\left(\frac{1}{q(x)}\right) + \int \left[r(x)q^2(x)\right]' p\left(\frac{1}{q(x)}\right) dx. \tag{4}$$

Now, following the idea used when solving Problem 18, we are led to consider the auxiliary function $g : I \to \mathbb{R}$, defined by

$$g(x) = \begin{cases} [r(x) \cdot q^2(x)]' p\left(\dfrac{1}{q(x)}\right), & \text{if } x \in I \backslash \{x_0\}, \\ 0, & \text{if } x = x_0. \end{cases}$$

Considering the assumptions a)–c) and e) we deduce that

$$\lim_{x \to x_0} [r(x) \cdot q^2(x)]' p\left(\frac{1}{q(x)}\right) = 0,$$

and hence $g$ is continuous at zero. This implies that $g$ is continuous on $I$ and hence it admits primitives on $I$. Let $G : I \to \mathbb{R}$ be one of its primitives. The relation (4) suggests that any primitive of $f$ will necessarily have the form

$$F(x) = \begin{cases} -r(x) \cdot q^2(x) \cdot p\left(\dfrac{1}{q(x)}\right) + G(x) + c_1, & \text{if } x \in I \backslash \{x_0\}, \\ c_2, & \text{if } x = x_0. \end{cases}$$

By imposing that $F$ be continuous at zero ($F$ must be differentiable and hence continuous on $I$) we get

$$c_2 = G(x_0) + c_1.$$

We shall prove that the function $F : I \to \mathbb{R}$ (obtained for $c_1 = 0$)

$$F(x) = \begin{cases} -r(x) \cdot q^2(x) \cdot p\left(\dfrac{1}{q(x)}\right) + G(x), & \text{if } x \in I \setminus \{x_0\}, \\ G(x_0), & \text{if } x = x_0, \end{cases} \tag{5}$$

is a primitive of $f$ on $I$.

Indeed, by the assumptions b)–c) we deduce that $F$ is differentiable at any $x \neq x_0$ and, moreover,

$$F'(x) = f(x).$$

For $x = x_0$ we have

$$\begin{aligned} F'(x_0) &= \lim_{x \to x_0} \frac{F(x) - F(x_0)}{x - x_0} \\ &= -\lim_{x \to x_0} \frac{r(x) \cdot q^2(x) \cdot p\left(\frac{1}{q(x)}\right)}{x - x_0} + \lim_{x \to x_0} \frac{G(x) - G(x_0)}{x - x_0}. \end{aligned}$$

By the assumptions c) and d) we get

$$\lim_{x \to x_0} \frac{r(x) \cdot q^2(x) \cdot p\left(\frac{1}{q(x)}\right)}{x - x_0} = 0$$

and

$$\lim_{x \to x_0} \frac{G(x) - G(x_0)}{x - x_0} = G'(x_0) = g(x_0) = 0,$$

respectively, which implies

$$F'(x_0) = 0 = f(x_0).$$ □

**Remark.** A function that admits primitives has also Darboux' property. Moreover, a function that has Darboux' property cannot have discontinuities of the second kind at which one lateral limit exists but is not finite. Therefore, in the framework stated by Theorem 1, the functions $f$ given by (3), for which at least one lateral limit at $x_0$ does not exist, form a class of discontinuous functions admitting primitives.

By considering some particular values of the functions $p, q$ and $r$ we recover the main steps of the route we have followed until we get, by successive explorations and investigations, our Theorem 1.

**Problem 18b.** *Let $n \in \mathbb{R}$ be a positive integer. The function*

$$f(x) = \begin{cases} \cos^n x \cdot \sin \dfrac{1}{x}, & x \in \mathbb{R}\backslash\{0\}, \\ 0, & x = 0, \end{cases}$$

*has primitives on $\mathbb{R}$.*

**Solution.** We take $I = J = (-\infty, \infty)$, $p(x) = -\cos x$, $q(x) = x$ and $r(x) = \cos^n x$, and apply Theorem 1. □

**Problem 18c.** *For $n \in \mathbb{N}$, the function*

$$f(x) = \begin{cases} \cos^n x \cdot \cos \dfrac{1}{x}, & x \in \mathbb{R}\backslash\{0\}, \\ 0, & x = 0, \end{cases}$$

*has primitives on $\mathbb{R}$.*

**Solution.** We take $I, J, q$ and $r$ as in Problem 18b, and $p(x) = \sin x$, and then apply Theorem 1. □

**Problem 18d.** *Let $a, b \in \mathbb{R}^*$ and $\alpha, t \in (0, \infty)$ with $\alpha < t$. The function*

$$f(x) = \begin{cases} \dfrac{a}{x^\alpha} \cos \dfrac{b}{x^t}, & x \in \mathbb{R}\backslash\{0\}, \\ 0, & x = 0, \end{cases}$$

*has primitives on $\mathbb{R}$.*

**Solution.** We apply Theorem 1 with $I, J$ as in Problem 18b, and $p(x) = \sin x$, $q(x) = \dfrac{x^t}{b}$, $r(x) = \dfrac{ab}{t} x^{1-\alpha-t}$. □

**Problem 18e.** *Let $n \in \mathbb{N}$. Show that the function $f : [0, \infty) \to \mathbb{R}$,*

$$f(x) = \begin{cases} \ln^n x \cdot \sin \dfrac{1}{x}, & x > 0, \\ 0, & x = 0, \end{cases}$$

*has primitives on $[0, \infty)$.*

**Solution.** We put $I = (0, \infty)$, $p$ and $q$ as in Problem 18b, $r(x) = \ln^n x$ (in this case, to check conditions d) and e) is a little bit more difficult), and then apply Theorem 1. □

**Problem 18f.** *If $p : \mathbb{R} \to \mathbb{R}$ is bounded and differentiable, then $f : \mathbb{R} \to \mathbb{R}$,*

$$f(x) = \begin{cases} p'\left(\dfrac{1}{x}\right), & x \neq 0, \\ 0, & x = 0, \end{cases}$$

*has primitives on $\mathbb{R}$.*

**Solution.** In this case $I = J = \mathbb{R}$, $q(x) = x$, $r(x) = 1$, $x \in I$. □

**Problem 18g.** *Let $q : \mathbb{R} \to \mathbb{R}$ be a function that has a unique zero $x_0 \in \mathbb{R}$ and fulfills the following conditions:*

1) *$q$ is differentiable and its first derivative is continuous on $\mathbb{R}\setminus\{x_0\}$;*

2) $\lim\limits_{x \to x_0} \dfrac{q^2(x)}{x - x_0} = 0$ *and* $\lim\limits_{x \to x_0} [q^2(x)]' = 0.$

*Then the function $f : \mathbb{R} \to \mathbb{R}$,*

$$f(x) = \begin{cases} \sin \dfrac{1}{q(x)}, & x \neq x_0, \\ 0, & x = x_0, \end{cases}$$

*has primitives on $\mathbb{R}$.*

**Solution.** We apply Theorem 1, taking $I, J$ and $p$ as in Problem 18b, and $r(x) = 1$. □

## 18.3 A Characterization of Functions Admitting Primitives

After solving the problems in the previous paragraphs, the following problem arises.

Let $f$ be a function of the form (3), which has the value $a$ ($a \in \mathbb{R}$) at the point $x = x_0$. For what numbers $a$ does the function $f$ have primitives?

An answer is given by

**Theorem 2.** *Let $a \in \mathbb{R}$. Let the conditions in Theorem 1 be satisfied. Then the function $f_a : I \to \mathbb{R}$,*

$$f_a(x) = \begin{cases} r(x) \cdot q'(x) \cdot p'\left(\dfrac{1}{q(x)}\right), & \text{for } x \in I \setminus \{x_0\}, \\ a, & \text{for } x = x_0, \end{cases} \tag{6}$$

*admits primitives on $I$ if and only if $a = 0$.*

**Proof.** The sufficiency follows by Theorem 1. In order to prove the necessity of condition $a = 0$, let us assume that there exists a primitive $F_a$ of the function $f_a$ on the interval $I$. Then, at least one of the intervals $(-\infty, x_0) \cap I$ and $(x_0, \infty) \cap I$ is not reduced to a single point. Let us denote it by $T$. The restrictions $F_a|_T$ and $F|_T$ are both primitives of the restriction $f_a|_T$, where by $F$ we denote the function given by (5). As any two primitives of the same function must differ by a constant on an interval, it results that there exists $c \in \mathbb{R}$ such that

$$F_a(x) = F(x) + c, \text{ for any } x \in T, \tag{7}$$

and since $F$ and $F_a$ are continuous at $x_0$ (being differentiable) we get that (7) holds for any $x \in T \cup \{x_0\}$. This implies that $F'_a(x_0) = F'(x_0)$ and hence

$$a = f_a(x_0) = F'_a(x_0) = F'(x_0) = 0. \qquad \square$$

In the following we indicate some important particular cases that led us to the statement of Theorem 2.

**Problem 18h.** Let $n \in \mathbb{N}^*$. *The function* $f_a : (-\pi, \pi) \to \mathbb{R}$,

$$f_a(x) = \begin{cases} \cos^n x \cdot \sin \dfrac{1}{\sin x}, & x \in (-\pi,\pi)\backslash\{x_0\}, \\ a, & x = 0, \end{cases}$$

*has primitives if and only if* $a = 0$.

**Solution.** We take $I = (-\pi, \pi)$, $J = (-\infty, \infty)$, $p(x) = -\cos x$, $q(x) = \sin x$ and $r(x) = \cos^{n-1} x$, and apply Theorem 2. $\square$

**Problem 18i.** *Let* $f : \mathbb{R} \to \mathbb{R}$ *be a differentiable function with* $f'$ *continuous,* $f(x) \neq 0$ *for all* $x \neq 0$ *and* $f(0) = 0$. *Find* $a \in \mathbb{R}$ *such that the function* $g : \mathbb{R} \to \mathbb{R}$ *given by*

$$g(x) = \begin{cases} f'(x) \cdot \sin \dfrac{1}{f(x)}, & x \neq 0, \\ a, & x = 0, \end{cases}$$

*has primitives on* $\mathbb{R}$.

**Solution.** We take $I, J$ and $p$ as in Problem 18b, $q = f$ and $r = 1_{\mathbb{R}}$, and apply Theorem 2. Condition d) is reduced here to $\lim\limits_{x \to 0} \dfrac{f^2(x)}{x} = 0$.

Since $f'$ is continuous, by l'Hospital's rule we have

$$\lim_{x \to 0} \frac{f(x)}{x} = \lim_{x \to 0} f'(x) = f'(0) \quad \text{(finite)}$$

and then

$$\lim_{x \to 0} \frac{f^2(x)}{x} = \lim_{x \to 0} f(x) \cdot \lim_{x \to 0} \frac{f(x)}{x} = f(0) \cdot f'(0) = 0.$$

Condition e) is also true, since

$$\lim_{x \to 0} [f^2(x)]' = \lim_{x \to 0} 2f(x) \cdot f'(x) = 0.$$

It is easy to check the other conditions in Theorem 2, hence $f$ has primitives if and only if $a = 0$. $\square$

## 18.4 Weakening the Assumptions

If we try to solve the next problem, we notice that it is not possible to use the results given in the previous paragraphs.

**Problem 18j.** Prove that *the function* $f : \mathbb{R} \to \mathbb{R}$,

$$f(x) = \begin{cases} \sin^{2n+1} \dfrac{1}{x}, & x \in \mathbb{R}\backslash\{0\}, \\ 0, & x = 0, \end{cases}$$

*has primitives on* $\mathbb{R}$.

Indeed, if we take $I = J = (-\infty, \infty)$, $p(x) = \cos x$, $q(x) = x$, $x \in \mathbb{R}$ and $r(x) = \sin^{2n} \dfrac{1}{x}$, $x \in \mathbb{R}\backslash\{0\}$, $n \in \mathbb{N}^*$, we find that conditions a)–d) in Theorem 1 are satisfied, but condition e) is not, because the limit

$$\lim_{x \to 0} \left( x^2 \sin^{2n} \frac{1}{x} \right)'$$

does not exist. Therefore, Theorem 1 cannot be applied here. To overcome these difficulties, we have to weaken the assumptions in Theorem 1 (and implicitly in Theorem 2 as well). One possible way is given by

**Theorem 3.** *The conclusion of Theorem* 1 *is valid if we replace c), d) and e) by the following three conditions:*

*c') p is differentiable on I;*

*d')* $\displaystyle\lim_{x \to x_0} \frac{r(x) \cdot q^2(x) \cdot p\left(\frac{1}{q(x)}\right)}{x - x_0} = 0;$

*e')* $\displaystyle\lim_{x \to x_0} [r(x) \cdot q^2(x)]' \cdot p\left(\frac{1}{q(x)}\right) = 0.$

The **proof** is similar to that of Theorem 1. □

Now we can solve Problem 18j. First we show by direct computations that the function $h : \mathbb{R} \to \mathbb{R}$,

$$h(x) = \sum (-1)^{k+1} \cdot \frac{1}{2k+1} C_n^k \cos^{2k+1} x,$$

is a primitive on $\mathbb{R}$ of the function $\sin^{2n+1} x$. Next, we use the fact that the function in Problem 18j can be written as

$$f(x) = \begin{cases} h'\left(\dfrac{1}{x}\right), & x \in \mathbb{R}\backslash\{0\}, \\ 0, & x = 0, \end{cases}$$

and then by applying Theorem 1 (or Theorem 3) we get that $f$ has primitives on $\mathbb{R}$. (The same conclusion could be obtained by using Problem 18f.) □

**Problem 18k.** *Let $f : \mathbb{R} \to \mathbb{R}$ be a differentiable function such that*

$$\lim_{x\to+\infty} \frac{f(x)}{x} = \lim_{x\to-\infty} \frac{f(x)}{x} = 0.$$

*Prove that the function $g : \mathbb{R} \to \mathbb{R}$,*

$$g(x) = \begin{cases} f'\left(\dfrac{1}{x}\right), & x \neq 0, \\ 0, & x = 0, \end{cases}$$

*has primitives on $\mathbb{R}$.*

**Solution.** We take $I = J = (-\infty, \infty)$, $p(x) = f(x)$, $q(x) = x$, $r(x) = 1$, $x \in \mathbb{R}$. The conditions in the hypothesis show that

$$\lim_{x\to 0} x\ f\left(\frac{1}{x}\right) = 0;$$

then we apply Theorem 3 (Theorem 1 cannot be applied in this case). □

The dual correspondent of Theorem 3 will be therefore

**Theorem 4.** *Let the assumptions of Theorem 3 be satisfied. Then the function $f_a$ given by (6) has primitives if and only if $a = 0$.*

If we take in Theorem 4, $I = (-\pi, \pi)$, $I = \mathbb{R}$, $p(x) = q(x) = \sin x$ and $r(x) = \cos^n x$, $x \in \mathbb{R}$, $n \in \mathbb{N}$, then we have just solved the next problem.

**Problem 18l.** *Let $n \in \mathbb{N}$. The function $f_a : (-\pi, \pi) \to \mathbb{R}$,*

$$f(x) = \begin{cases} \cos^n x \cos \dfrac{1}{\sin x}, & x \in (-\pi, \pi)\setminus\{0\}, \\ a, & x = 0, \end{cases}$$

*has primitives if and only if $a = 0$.*

## 18.5 The Function Has More Discontinuity Points

All the functions we considered so far had a unique point of discontinuity, generally $x_0 = 0$ (when $x_0 \neq 0$ we can reduce the problem by a simple transformation to the case $x_0 = 0$). It is then natural to wonder if the results are still valid when the set of discontinuity points of $f$ is richer. We get a more general result stated in

**Theorem 5.** *Let $p, q$ and $r$ fulfil the following conditions:*

*a) the set $A = \{x \in I \mid q(x) = 0\}$ is at most countable;*

*b) $q$ and $r$ are differentiable on $I\setminus A$ and their derivatives are continuous on $I\setminus A$;*

*c) $p$ is bounded and differentiable on $I$;*

*d)* $\lim_{x \to a} \frac{r(x) \cdot q^2(x)}{x - a} = 0$ *for any* $a \in A$;

*e)* $\lim_{x \to a} [r(x) \cdot q^2(x)]' = 0$ *for any* $a \in A$.

*Then the function* $f : I \to \mathbb{R}$,

$$f(x) = \begin{cases} r(x) \cdot q'(x) \cdot p'\left(\frac{1}{q(x)}\right), & \text{for } x \in I \setminus A, \\ 0, & \text{for } x \in A, \end{cases}$$

*has primitives on* $I$.

Now we illustrate how to use Theorem 5 for solving the next problem.

**Problem 18m.** *Let* $A = \left\{x \in \mathbb{R} | x = \frac{1}{k},\ k \in \mathbb{Z}^*\right\} \cup \{0\}$. *Prove that the function* $f : \mathbb{R} \to \mathbb{R}$ *given by*

$$f(x) = \begin{cases} \cos\frac{\pi}{x} \sin\frac{1}{x^2 \sin\frac{\pi}{x}}, & x \in \mathbb{R} \setminus A, \\ 0, & x \in A, \end{cases}$$

*has primitives on* $\mathbb{R}$.

**Solution.** We take $I, J$ and $p$ as in Problem 18b, $r(x) = 1$ and $q(x) = x^2 \sin\frac{\pi}{x}$, and by applying Theorem 5 we obtain that the function

$$h(x) = \begin{cases} \left(2x \sin\frac{\pi}{x} - \pi \cos\frac{\pi}{x}\right) \sin\frac{1}{x^2 \sin\frac{\pi}{x}}, & x \in \mathbb{R} \setminus A, \\ 0, & x \in A, \end{cases}$$

has primitives. On the other hand, the function $g : \mathbb{R} \to \mathbb{R}$,

$$g(x) = \begin{cases} 2x \sin\frac{\pi}{x} \sin\frac{1}{x^2 \sin\frac{\pi}{x}}, & x \in \mathbb{R} \setminus A, \\ 0, & x \in A, \end{cases}$$

is continuous on $\mathbb{R}$ and hence it has primitives. As

$$f = \frac{1}{\pi}(g - h),$$

it results that $f$ admits primitives, too. □

At the end of this section we recommend that readers explore by themselves other directions of possible extensions for Problem 18, by exploiting the periodicity of the involved functions, see [1], [10] and [12].

## 18.6 Bibliographical Comments

Problem 18 belongs to the mathematical "folklore". It was difficult to establish its origin. The first source we could find in the Romanian literature is the textbook [5], p. 14, exercise III.4. There we found also Problem 18a as exercise III.3.

The content of paragraph 18.2 reproduces practically our paper [4], slightly adapted. The same results have been inserted in the book [2] written by V. Arsinte. The case of many discontinuity points (actually a set at most countable) has been studied in our paper [3] where the results are presented in a very interesting fixed point formulation.

Problems 18b–18e are taken from [4]. They are generalizations of several problems from *Gazeta Matematica* or from [5]. Thus, from Problem 18b we obtain for $n = 0$ Problem 18, while from Problem 18c we obtain Problem 18a; for $\alpha = \frac{1}{2n+1}$, $t = 1$ and $b = a$, from Problem 18d we obtain the problem 20530 in *Gazeta Matematica* no. 8/1985, author L. Coroian. For $n = 1$, from Problem 18e we obtain the problem 19840* in *Gazeta Matematica* no. 8/1983, author H. Wilk.

Problem 18h is the problem 19601* in *Gazeta Matematica* no. 2/1983, author S. Mihaescu, while Problem 18i is Problem 20501* from *Gazeta Matematica* no. 7/1987, author V. Blendea and G. Blendea.

Problem 18j is a partial statement of Problem 2 given at the county round of the National Mathematical Olympiad 1982, see *Gazeta Matematica* no. 6/1982, pp. 220–221.

Problem 18k is part a) of Problem 3 given at the county round of the National Mathematical Olympiad in 1983, see *Gazeta Matematica* no. 7/1983, p. 276, author being L. Panaitopol. Part b) of the same problem can be obtained from Problem 18d, for $a = b = 1$, $\alpha = \frac{1}{3}$ and $t = 1$.

Problem 18l, which can be obtained as a particular case ($n = 0$) of application 11 in [4], is in fact the problem 20335 from *Gazeta Matematica* no. 1/1985, author V. Berinde.

Theorem 5 in this section is Theorem 2.1 of [3], while Problem 18m represents problem C:559 in *Gazeta Matematica* no. 12/1985, author O. Cornea.

## 18.7 References

1. **D. Andrica,** *On a Large Class of Functions Admitting Primitives.* (Romanian), Gazeta Matematica (Serie metodica), **4 (75)** (1986), 169–176

2. **V. Arsinte,** *Elementary Problems of Integral Calculus.* (Romanian), Editura Universitatii Bucuresti, 1995

3. **V. Berinde,** *A Fixed Point Characterization of the Points of Discontinuity of a Derivative*, Seminar on Fixed Points Theory, **3** (1985), 9–14

4. **V. Berinde,** *A Class of Discontinuous Functions Admitting Primitives.* (Romanian), Gazeta Matematica, **94** (1989), no. 6, 214–220

5. **N. Boboc, I. Colojoara,** *Elements of Mathematical Analysis. Textbook for XII Form.* (Romanian), Editura Didactica si Pedagogica, 1982

6. **M. Ganga,** *Themes and Problems of Mathematics.* (Romanian), Editura Tehnica, 1991

7. **O. Konnerth,** *Typical Errors in Learning Mathematical Analysis.* (Romanian), Editura Dacia, 1982

8. **I. Muntean,** *On the Existence of Primitives for and Integrability of Continuous Functions (I).* (Romanian), Gazeta Matematica (Serie metodica), **2 (71)** (1981), no. 2-3, 60–67

9. **I. Muntean,** *On the Existence of Primitives for and Integrability of Continuous Functions (II).* (Romanian), Gazeta Matematica (Serie metodica), **2 (71)** (1981), no. 4, 165–175

10. **E. Paltanea,** *A Class of Functions Admitting Primitives.* (Romanian), Gazeta Matematica (Serie metodica), **3 (72)** (1982), no. 3-4 , 137–146

11. **L. Petracovici,** *On the Existence of the Primitives of Real Functions.* (Romanian), Lucr. Semin. Creativ. Mat., **1** (1991-1992), 105–122

12. **F. Vulpescu-Jalea,** *On the Functions Admitting Primitives.* (Romanian), Gazeta Matematica (Serie metodica), **2 (71)** (1981), no. 3-4, 159–166

13. **I.L. Dinulescu, G. Nedelea,** *The Unitary Treatment of Some Problems Regarding the Existence of Primitives of the Functions.* (Romanian), Gazeta Matematica, **100** (1995), no. 8, 349–361.

14. **Bourbaki, N.,** *Éléments de mathématique.* Fasc. IX. Livre IV: Fonctions d'une variable réelle. Thorie élémentaire. Chap. 1: Dérivées. Chap. 2: Primitives et intégrales. Chap. 3: Fonctions élémentaires. Actualites scientifiques et industrielles. No.1074. Hermann, Paris, 1968

# Theme 19

# On Two Classes of Inequalities

In this section we will approach two classes of inequalities, somehow related to one another, as they represent segments of the Riemann series $\sum_{n\geq 1} n^{-\alpha}$, $\alpha \geq 1$.

## 19.1 First Source Problem: Telescopic Sums

**Problem 1.** *Prove that for any $n \in \mathbb{N}^*$,*

$$1 + \frac{1}{2\sqrt{2}} + \frac{1}{3\sqrt{3}} + \ldots + \frac{1}{n\sqrt{n}} \geq \frac{2n}{n+1}.$$

**Solution.** As $\sqrt{k} = \sqrt{k \cdot 1} \leq \dfrac{k+1}{2}$, $\forall\, k \geq 1$, we deduce that

$$\frac{1}{\sqrt{k}} \geq \frac{2}{k+1} \text{ and so } \frac{1}{k\sqrt{k}} \geq \frac{2}{k(k+1)} = 2\left(\frac{1}{k} - \frac{1}{k+1}\right).$$

Adding up we have

$$1 + \frac{1}{2\sqrt{2}} + \frac{1}{3\sqrt{3}} + \ldots + \frac{1}{n\sqrt{n}}$$
$$\geq 2\left(\frac{1}{1} - \frac{1}{2} + \frac{1}{2} - \frac{1}{3} + \ldots + \frac{1}{n} - \frac{1}{n+1}\right) = \frac{2n}{n+1}$$

which is exactly the inequality in the problem. □

**Remark.** The basic idea for solving this problem was using the means inequality in order to obtain a "telescopic" sum, that is a sum that can be "compressed" by cancelling the like terms and leaving thus only the first and the last terms of the sum. This method is often used to get bounds for the sums in the Riemann series. For example in case $\alpha = 2$ we have

$$\frac{1}{k^2} > \frac{1}{k(k+1)} = \frac{1}{k} - \frac{1}{k+1}, \quad k \geq 1$$

and

$$\frac{1}{k^2} < \frac{1}{(k-1)k} = \frac{1}{k-1} - \frac{1}{k}, \quad k \geq 2.$$

Using this idea, we can similarly solve related problems and propose new problems as well. We will now present some of these.

**Problem 19a.** *Prove that if $n \in \mathbb{N}$, $n \geq 2$, then*

$$1 + \frac{1}{2^2} + \ldots + \frac{1}{n^2} > \frac{3n}{2n+1}.$$

**Solution.** We have

$$\frac{1}{3k^2} \geq \frac{1}{4k^2-1} = \frac{k}{2k+1} - \frac{k-1}{2k-1}$$

and so

$$\sum_{k=1}^{n} \frac{1}{3k^2} > \sum_{k=1}^{n} \left( \frac{k}{2k+1} - \frac{k-1}{2k-1} \right) = \frac{n}{2n+1}.$$

□

**Problem 19b.** *Prove that for any $n \in \mathbb{N}$, $n \geq 1$, we have*

$$\frac{1}{3} + \frac{1}{2 \cdot 4\sqrt{2}} + \frac{1}{3 \cdot 5\sqrt{3}} + \ldots + \frac{1}{n(n+2)\sqrt{n}} \geq \frac{n(n+3)}{2(n+1)(n+2)}.$$

**Hint.** We use the relations

$$\frac{1}{k(k+2)\sqrt{k}} \geq \frac{2}{k(k+1)(k+2)} = \frac{1}{k(k+1)} - \frac{1}{(k+1)(k+2)}.$$

**Problem 19c.** *For $n \in \mathbb{N}$, $n \geq 1$, we have*

$$\sum_{k=1}^{n} \frac{1}{(k-1)!\sqrt{k}} \geq 2\left[1 - \frac{1}{(n+1)!}\right].$$

**Hint.**

$$\begin{aligned} \frac{1}{(k-1)!\sqrt{k}} &\geq \frac{2}{(k-1)!(k+1)} = \frac{2k}{(k+1)!} = 2 \cdot \frac{k+1-1}{(k+1)!} \\ &= 2\left(\frac{1}{k!} - \frac{1}{(k+1)!}\right). \end{aligned}$$

**Problem 19d.** *Prove that*

$$\sum_{k=1}^{n} \frac{1}{(2k-1)\sqrt{k^2+k}} \geq \frac{2n}{2n+1}.$$

**Hint.** Since

$$\sqrt{k^2+k} = \sqrt{k(k+1)} \leq \frac{k+k+1}{2} = \frac{2k+1}{2}$$

we get the telescopic sum

$$\sum_{k=1}^{n}\left(\frac{1}{2k-1}-\frac{1}{2k+1}\right)=1-\frac{1}{2n+1}=\frac{2n}{2n+1}. \qquad \square$$

We recommend that readers follow the same recipe and propose themselves new problems using other inequalities than those presented here.

## 19.2 Second Source Problem: Sequences of the Harmonic Series

**Problem 19e.** *Prove that the following inequalities hold for* $n \in \mathbb{N}^*$:

$$\frac{1}{5n+1}+\frac{1}{5n+2}+\ldots+\frac{1}{25n}>\frac{6}{5}, \tag{1}$$

$$\frac{1}{7n+1}+\frac{1}{7n+2}+\ldots+\frac{1}{49n}>\frac{11}{7}. \tag{2}$$

To simplify the exposition of the general method used for solving this type of inequalities, we will first solve the following simpler problem.

**Problem 19f.** *For* $n \in \mathbb{N}^*$, $n \geq 2$, *we have*

$$\frac{1}{2n+1}+\frac{1}{2n+2}+\ldots+\frac{1}{4n}>\frac{7}{12}. \tag{3}$$

**Solution.** Inequality (3) is true for $n = 2$, as it becomes

$$\frac{1}{5}+\frac{1}{6}+\frac{1}{7}+\frac{1}{8}>\frac{7}{12} \Leftrightarrow \frac{533}{210}>\frac{7}{3}.$$

We suppose that (3) holds for $n = k$, i.e.,

$$\frac{1}{2k+1}+\frac{1}{2k+2}+\ldots+\frac{1}{4k}>\frac{7}{12}, \tag{4}$$

and now let us prove that (3) holds for $n = k + 1$ as well, i.e.,

$$\frac{1}{2k+3}+\frac{1}{2k+4}+\ldots+\frac{1}{4k+2}+\frac{1}{4k+3}+\frac{1}{4k+4}>\frac{7}{12}. \tag{5}$$

To be able to use the relation (4), supposedly true, we complete the missing terms and eliminate those that do not appear in (5), so that from (4) we get an inequality containing the left-hand side of (5). This means adding to both sides of (4) the quantity

$$E=\frac{1}{4k+1}+\frac{1}{4k+2}+\frac{1}{4k+3}+\frac{1}{4k+4}-\frac{1}{2k+1}-\frac{1}{2k+2}.$$

Thus we get

$$\frac{1}{2k+3}+\frac{1}{2k+4}+\ldots+\frac{1}{4k+4}\geq\frac{7}{12}+E, \tag{6}$$

which we know to be true.

To prove, using (6), that (5) is true as well, it will be enough to prove that

$$\frac{7}{12}+E\geq\frac{7}{12},\text{ i.e., } E\geq 0, \tag{7}$$

which is equivalent to

$$\frac{1}{4k+1}+\frac{1}{4k+2}+\frac{1}{4k+3}+\frac{1}{4k+4}-\frac{1}{2k+1}-\frac{1}{2k+2}\geq 0. \tag{8}$$

But $\frac{1}{4k+2}-\frac{1}{2k+1}=-\frac{1}{4k+2}$ and $\frac{1}{4k+4}-\frac{1}{2k+2}=-\frac{1}{4k+4}$, and so (8) becomes

$$\frac{1}{4k+1}+\frac{1}{4k+3}\geq\frac{1}{4k+2}+\frac{1}{4k+4},$$

which is obviously true.

Therefore (7) is true and so, having in mind (6), it results that (5) is true as well. So we have just proved that if (3) is true for $n=k$, then it is also true for $n=k+1$.

By the principle of mathematical induction this means that (3) is true for any $n\in\mathbb{N}^*$. □

**Remarks.** 1) If, instead of (3), the problem asked to prove the inequality

$$\frac{1}{3n+1}+\frac{1}{3n+2}+\ldots+\frac{1}{6n}>\frac{3}{5} \tag{9}$$

or

$$\frac{1}{4n+1}+\frac{1}{4n+2}+\ldots+\frac{1}{16n}>\frac{5}{8} \tag{10}$$

or even the inequality more rich in terms

$$\frac{1}{3n+1}+\frac{1}{3n+2}+\ldots+\frac{1}{9n}>1,$$

then the expression $E$ would have more terms and it would have been very difficult to prove by direct computations that $E\geq 0$ holds. For instance, the inequality $E\geq 0$ would have, in the case of (9), $6+3=9$ terms; in the case of (10) it would have $8+4=12$ terms; in the case of (1), it would have $25+5=30$ terms, while in the case of (2), $49+7=56$ terms.

2) Anyhow, we notice that the inequality (8) can be easily obtained from

$$\frac{1}{4n+1}+\frac{1}{4n+2}>2\cdot\frac{1}{4k+2}=\frac{1}{2k+1}$$

and

$$\frac{1}{4n+3}+\frac{1}{4n+4}>2\cdot\frac{1}{4k+4}=\frac{1}{2k+2},$$

without performing any other computations.

3) To prove that (9) holds, we must prove that

$$\frac{1}{6k+1}+\frac{1}{6k+2}+\ldots+\frac{1}{6k+6}-\frac{1}{3k+1}-\frac{1}{3k+2}-\frac{1}{3k+3}\geq 0.$$

Here we will use the fact that

$$\frac{1}{6k+1}+\frac{1}{6k+2}\geq\frac{2}{6k+2}=\frac{1}{3k+1},\ \text{and so on.}$$

4) The technique of grouping terms, by 5 in equation (1) and by 7 in (2), is now a very useful tool in applying the method of induction in order to prove the inequalities of the type described in this paragraph.

5) If we carefully analyze this procedure, we notice that in fact the method of grouping terms can be applied directly in order to prove the inequalities, without using the induction method, as shown in the following

**Solution of Problem 19e.** In the left-hand side of the inequality (1) we have $25n - 5n = 20n$ terms, which we group by $5n$, getting

$$\frac{1}{5n+1}+\frac{1}{5n+2}+\ldots+\frac{1}{10n}>5n\cdot\frac{1}{10n}=\frac{1}{2},$$

$$\frac{1}{10n+1}+\frac{1}{10n+2}+\ldots+\frac{1}{15n}>5n\cdot\frac{1}{15n}=\frac{1}{3},$$

$$\frac{1}{15n+1}+\frac{1}{15n+2}+\ldots+\frac{1}{20n}>5n\cdot\frac{1}{20n}=\frac{1}{4}$$

and, respectively,

$$\frac{1}{20n+1}+\frac{1}{20n+2}+\ldots+\frac{1}{25n}>5n\cdot\frac{1}{25n}=\frac{1}{5},$$

which added up, side by side, give

$$\frac{1}{5n+1}+\frac{1}{5n+2}+\ldots+\frac{1}{25n}>\frac{1}{2}+\frac{1}{3}+\frac{1}{4}+\frac{1}{5}.$$

The inequality (1) results now from the fact that

$$\frac{1}{2}+\frac{1}{3}+\frac{1}{4}+\frac{1}{5}=\frac{77}{60}\ \text{and}\ \frac{77}{60}>\frac{6}{5}\ (\Leftrightarrow\ 385>360).$$

For the inequality (2), the method is similar. □

**Remark.** From the previous remarks, we deduce that the inequality (1) can be strengthened as follows.

**Problem 19g.** *For $n \in \mathbb{N}^*$, we have*

$$\frac{1}{5n+1}+\frac{1}{5n+2}+\ldots+\frac{1}{25n}>\frac{77}{60}.$$

We can obtain a stronger inequality if we group the terms in the left-hand side by $n$, and so in the right-hand side we will have

$$\frac{1}{6}+\frac{1}{7}+\frac{1}{8}+\ldots+\frac{1}{25}. \tag{11}$$

We can evaluate this sum more exactly and we will get a stronger inequality than the one in Problem 19g, as using the same technique for evaluating the sum (11) by grouping each 5 terms, we get

$$\frac{1}{6}+\frac{1}{7}+\ldots+\frac{1}{25}>\frac{1}{2}+\frac{1}{3}+\frac{1}{4}+\frac{1}{5}=\frac{77}{60}.$$

Thus we can obtain new inequalities. For instance, starting off from (10) and grouping by $n$ terms, we get

$$\begin{aligned}\frac{1}{4n+1}+\frac{1}{4n+2}+\ldots+\frac{1}{16n} &> n\cdot\left(\frac{1}{5n}+\frac{1}{6n}+\ldots+\frac{1}{16n}\right)\\ &= \frac{1}{5}+\frac{1}{6}+\ldots+\frac{1}{16}\end{aligned}$$

and then, similarly,

$$\frac{1}{4}+\frac{1}{6}+\ldots+\frac{1}{16}>4\left(\frac{1}{8}+\frac{1}{12}+\frac{1}{16}\right)=\frac{1}{2}+\frac{1}{3}+\frac{1}{4}=\frac{13}{12}.$$

Thus we have obtained a new inequality, comprised in

**Problem 19h.** *Prove that for $n \in \mathbb{N}^*$, we have*

$$\frac{1}{4n+1}+\frac{1}{4n+2}+\ldots+\frac{1}{16n}>\frac{13}{12}.$$

By taking fewer members of the sum, we get the inequalities

$$\begin{aligned}\frac{1}{4n+1}+\frac{1}{4n+2}+\ldots+\frac{1}{12n}&>\frac{5}{6},\\ \frac{1}{5n+1}+\frac{1}{5n+2}+\ldots+\frac{1}{10n}&>\frac{3}{5},\end{aligned} \tag{12}$$

and so on. We invite readers to extend in the same manner the inequalities in

**Problem 19i.** *Prove that*

$$\frac{1}{5n+1}+\frac{1}{5n+2}+\ldots+\frac{1}{10n}>\frac{10n}{15n+1},\quad n \in \mathbb{N}^*; \tag{13}$$

$$\frac{1}{pn+1}+\frac{1}{pn+2}+\frac{1}{pn+3}+\ldots+\frac{1}{2pn}>\frac{2pn}{3pn+1},\quad n,p \in \mathbb{N}^*. \tag{14}$$

## 19.3 Bibliographical Comments

Problem 19 was published in *Gazeta Matematica* no. 3/1994, as problem E: 10704 by Marian Ioan Popa; Problem 19a is taken from [2], p. 42, problem 2.91. The other problems from Section 19.2 are *variations* on the same theme.

Problem 19e was published in *Revista Matematica a Elevilor din Timisoara* as problem no. 2078 in 1976. It is included in [4] entirely, and part of it in [2] without a solution, but followed by the hint: "it should be proved using the method of induction".

Section 19.2 is basically a version of our article [1]. The inequalities (9) and (10) are the subject of problems 10.39 and 10.40, p. 153, from [2]. The inequalities (12) are part of problem 18748 from *Gazeta Matematica* no. 5/1981, author I. Atanasiu, while (13) and (14) appear in problem C:188, *Gazeta Matematica*, no. 2-3/1981, author Liliana Niculescu.

## 19.4 References

1. **V. Berinde,** *Demonstrating Some Inequalities with and without Induction.* Lucr. Semin. Creativ. Mat., **6** (1995–1997), 33–40

2. **C. Militaru,** *Exercises and Problems for Secondary Schools and Entrance Examinations for University.* (Romanian), Editura Alux, 1992

3. **L. Pirsan, C. Ionescu-Tiu,** *Mathematics Problems for Secondary-School Students in Forms XI and XII.* (Romanian), Editura Facla, 1979

4. **L. Pirsan, C. Lazanu,** *Problems of Algebra and Trigonometry for Secondary School Students in Forms IX and X.* (Romanian), Editura Facla, 1983

# Theme 20

# Another Problem of Geometric Construction

## 20.1 Source Problem. Some Exploratory Directions

**Problem 20.** *Let $M$ be the midpoint of the side $BC$ in the triangle $ABC$. Construct, using ruler and compass, the points $E \in AB$ and $F \in AC$ such that $EF \| BC$ and $EM \perp FM$.*

**Solution.** We suppose the problem already solved and let $E \in AB$, $F \in AC$ be such that $EF \| BC$ and $EM \perp FM$ (see Figure 10).

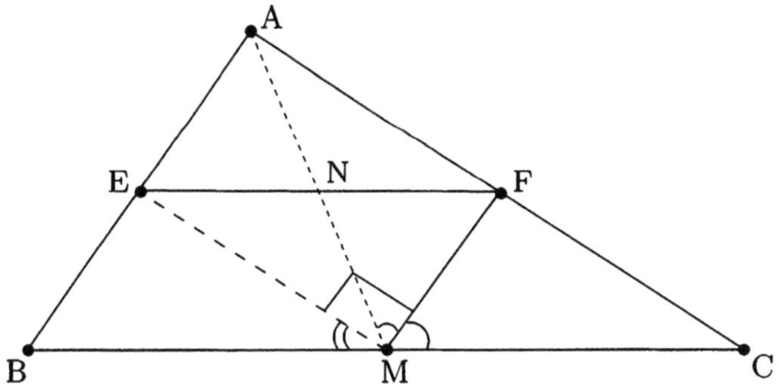

Figure 10

Since $m(\measuredangle EMF) = 90^0$, it results that

$$m(\measuredangle EMB) + m(\measuredangle FMC) = 90^0. \tag{1}$$

As $EF \| BC$, we get by Thales' theorem that

$$\frac{AE}{AB} = \frac{EN}{BM} = \frac{AN}{AM}, \tag{2}$$

where $\{N\} = AM \cap EF$.

$N$ is obviously the midpoint of $[EF]$, so $EMF$ being a right triangle implies (the median = a half of the hypothenuse) that:

$$MN = EN = NF. \tag{3}$$

By (2) we now get

$$\frac{AE}{AB - AE} = \frac{AN}{AM - AN} = \frac{EN}{BM - EN},$$

thus by the first equality we have

$$\frac{AE}{EB} = \frac{AN}{NM}. \tag{4}$$

By (2) and (3) we get

$$\frac{AN}{AM} = \frac{NM}{BM} \Leftrightarrow \frac{AN}{NM} = \frac{AM}{BM} \tag{5}$$

and by (4) and (5) we now get $\frac{AE}{EB} = \frac{AM}{BM}$. Using the bisectrix theorem ([3], p. 39) we obtain that $ME$ is the bisectrix of the angle $\measuredangle AMB$. Similarly we show that $MF$ is the bisectrix of $\measuredangle AMC$.

The construction follows immediately. We use the bisectrix theorem and construct on $AB$ a point $E$ which divides $[AB]$ in segments proportional to the sides $BM$ and $AM$ of the triangle $ABM$ ($EM$ is the bisectrix) by the known method, see Figure 11.

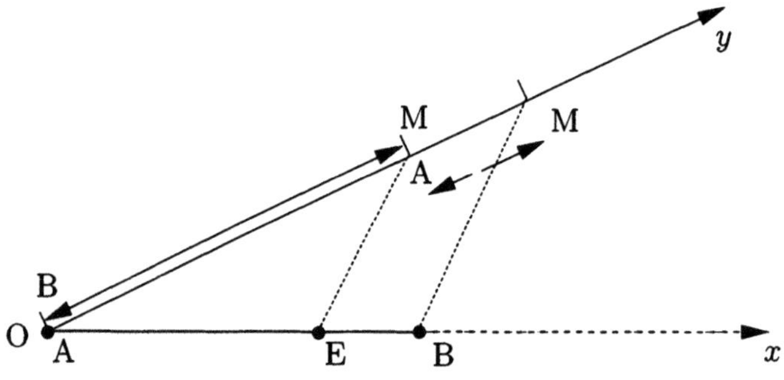


Figure 11

The solution of the problem is unique. Indeed, in spite of the fact that by hypothesis $E \in AB$ and $F \in AC$, we can have only $E \in [AB]$ and $F \in [AC]$. Let us assume that $E_1$ and $F_1$ were situated on the extensions of the sides $[AB]$ and $[AC]$, respectively. Then we could not apply the bisectrix theorem to obtain that $E_1F_1 \| BC$ (see Figure 12).

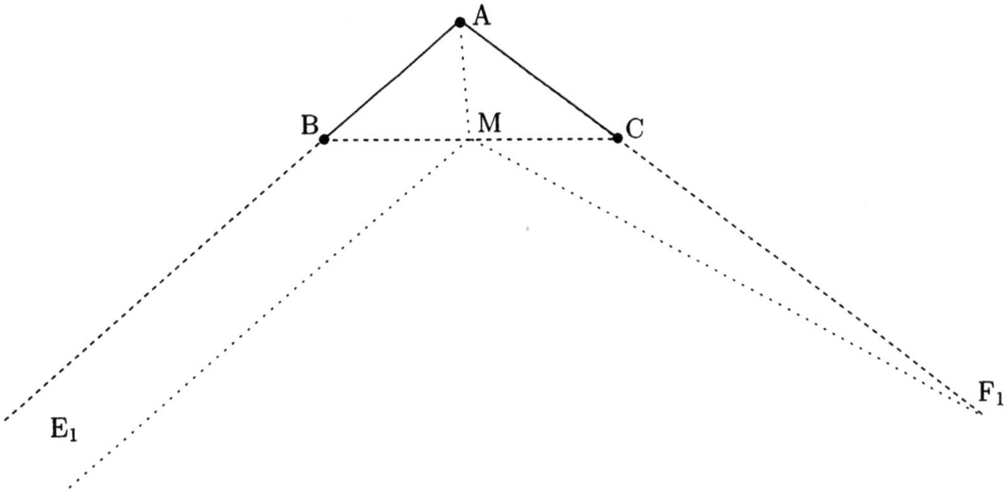


Figure 12

**Remark.** If we read carefully the solution of Problem 20 we notice that we have exploited actually only the fact that $M$ is the midpoint of the side $[BC]$, a condition that has led us to the conclusion that $EM$ and $FM$ are the bisectrices of the angles $\measuredangle ABM$ and $\measuredangle AMC$, respectively.

The following question naturally arises: are $ME$ and $MF$ still bisectrices, even if $M$ is another important point on $BC$ as, for example, the foot of the altitude on $BC$ in the triangle $ABC$? In order to answer this question we first solve the following problem.

**Problem 20a.** *Let $D$ be the intersection of the altitude drawn from $A$ in the acute angle triangle $ABC$. Construct the points $E \in (AB)$ and $F \in (AC)$ such that $EF \| BC$ and $ED \perp FD$.*

**Solution.** We consider, as previously, the problem already solved and let $E \in (AB)$, $F \in (AC)$ be the required points (see Figure 13).

On $[BC]$ as diameter we construct a semicircle situated in the half-plane determined by $BC$, that does not contain the point $A$.

We extend $AD$ until it intersects the semicircle at $P$ (since $ABC$ is an acute angle triangle, the point $D$ belongs to $(BC)$ and hence $AD$ always intersects the semicircle).

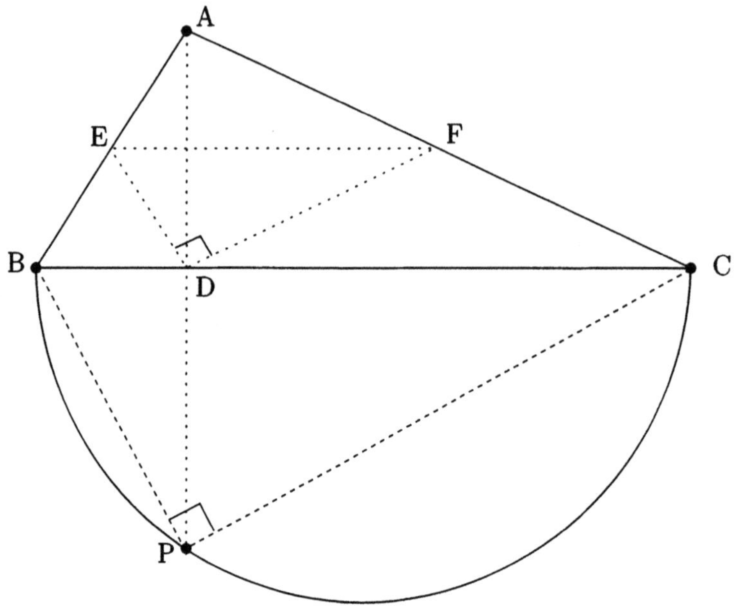


Figure 13

It is very clear that the angle $\measuredangle BPC$ is a right angle at $P$ and, since $EF \| BC$, $DE \| BP$, we obtain that $DF \| PC$ as well. The required construction follows immediately.

a) We construct the semicircle of diameter $[BC]$ as above;

b) We extend $AD$ (where $AD \perp BC$), until it intersects the semicircle at the point $P$, let us say;

c) We draw by $D$ a parallel to $BP$, which intersects $AB$ in $E$, and a parallel to $PC$ which intersects $AC$ at $F$. The points $E$ and $F$ thus obtained are just those required by the problem. Indeed, $\sphericalangle EDF \equiv \sphericalangle BPC$ (angles with parallel sides and $m(\sphericalangle BPC) = 90^0$).
By applying Thales' theorem in triangles $ABP$ and $APC$ we get

$$\frac{AE}{AB} = \frac{AD}{AP} \text{ and } \frac{AF}{AC} = \frac{AD}{AP}, \text{ respectively.}$$

Therefore $\dfrac{AE}{AB} = \dfrac{AF}{AC}$ and, by the reciprocal of Thales' theorem, we have that $EF \| BC$.

**Remark.** It is easy to notice that Problem 20a contains two additional restrictions compared to Problem 20.

**Restriction 1.** "$ABC$ is an acute angle triangle";
**Restriction 2.** "$E \in (AB)$ and $F \in (AC)$".

The first one is necessary, as we have seen, to ensure that $D$ belongs to the interior of $[BC]$. The second restriction excludes the possibility of having another pair of points $(E, F)$ such that $m(\sphericalangle EDF) = 90^0$. We shall discuss these aspects again after we solve

**Problem 20b.** *Let $ABC$ be a triangle and $M \in (BC)$, arbitrary. Construct the points $E \in (AB)$, $F \in (AC)$ such that $EF \| BC$ and $EM \perp MF$.*

**Solution.** We follow the same steps as in the previous problem. For any triangle $ABC$, even with an obtuse angle, the line $AM$ will always intersect the semicircle of diameter $[BC]$, situated in the half-plane determined by $BC$ and which does not contain the point $A$, at a unique point $P$.

The rest of the construction runs as in Problem 20a.

**Remarks.** 1) It is now clear that the assumption "$ABC$ is an acute angle triangle" which appears in [1], the problem G1, is not necessary to perform the construction. That assumption was due to the special case when $D$ is the foot of the altitude of the triangle $ABC$ which passes through $A$;

2) It is now easy to notice that we can remove the assumption that $\sphericalangle EMF$ is a right angle, as in the next problem.

**Problem 20c.** *Let $M$ be a point on the side $BC$ of the triangle $ABC$. Construct the points $E \in (AB)$ and $F \in (AC)$ such that $m(\sphericalangle EMF) = \alpha$ (given) and $EF \| BC$.*

**Solution.** We construct the arc capable of the (inscribed) angle $\alpha$ that passes through the points $B$ and $C$, situated in the half-plane determined by $BC$ that does not contain $A$ (see [3], pp. 61–62, for example). Then

$$m(\sphericalangle BPC) = \alpha$$

and the rest of the proof follows as in Problem 20b.

**Remark.** Now, using the method for solving Problem 20a, also extended to Problem 20b, we notice that in the statements of Problems 20a–20c we can ask only $E \in AB$ and $F \in AC$, since the obtained answer will force $E \in (AB)$ and $F \in (AC)$.

If we notice that all the considerations in this section are in the plane, we can extend all the problems in space, too, following the next sample problem.

**Problem 20d.** *Let $ABCD$ be a tetrahedron and $G$ be the centroid of the triangle $ABC$. Construct the points $M \in (DBC)$, $N \in (DAC)$ and $P \in (DAB)$ such that the tetrahedron $MNPG$ has three right angles with respect to $G$ and $(MNP) \| (ABC)$.*

We leave the solution of this problem to the readers and also suggest that they consider the other variations on this idea, for the case when $G$ is replaced by other important points in the triangle $ABC$.

## 20.2 Bibliographical Comments

Problem 20 was published in *Gazeta Matematica* no. 2/1995 as problem 23176, author T. Zvonaru.

The first results of the explorations done around this problem have been published in the paper [2]. Similar results are obtained in [1] as well.

Problem 20a was proposed by T. Zvonaru in *Gazeta Matematica* no. 8/1985 as problem 23322.

## 20.3 References

1. **D. Barbosu, C. Cioclu,** *On a Problem of Geometric Construction.* (Romanian), Gazeta Matematica, **105** (2000), no. 1, 20–21

2. **V. Berinde,** *A Natural Way to Initiate Students in Mathematics Research: Creative Problems Solving Sessions.* Heuristics and Didactics of Exact Sciences (Donetsk), no. 2 (1997), 30–32

3. **A. Cota et al.,** *Mathematics. Geometry and Trigonometry. Textbook for IX Form.* (Romanian), Editura Didactica si Pedagogica, Bucuresti, 1993

# Theme 21

# How Can We Discover New Problems by Means of the Computer

## 21.1 Introduction

In this section we want to show that using the computer we can get the primary data necessary for discovering new problems.

Using the author's experience gathered in the papers [1]–[5], we maintain that it is possible to obtain new qualitative results, by quantitative processing with a computer. Therefore, the basic idea of this section is not only that the computer can be used for solving purely theoretical problems, but that we can also discover new problems which the computer proposes to the theoretical mathematics.

## 21.2 First Source Problem

**Problem 21.1.** *Find natural numbers $x, y, z, t, u$, so that the following conditions are simultaneously satisfied:*

$$1)\ x \neq 2; \qquad 2)\ x < y < z < t < u; \qquad 3)\ \frac{1}{x} + \frac{1}{y} + \frac{1}{z} + \frac{1}{t} + \frac{1}{u} = 1.$$

**Solution.** From condition 1) we deduce that $x \geq 3$ (we cannot have $x = 1$). From condition 2) we have $\frac{1}{x} > \frac{1}{y}, \frac{1}{x} > \frac{1}{z}, \frac{1}{x} > \frac{1}{t}, \frac{1}{x} > \frac{1}{u}$ and substituting in 3) we obtain $1 = \frac{1}{x} + \frac{1}{y} + \frac{1}{z} + \frac{1}{t} + \frac{1}{u} < \frac{5}{x}$, from which it results that $x < 5$. Therefore $x \in \{3, 4\}$.

Let $x = 3$. From 3) it results that

$$\frac{1}{y}+\frac{1}{z}+\frac{1}{t}+\frac{1}{u}=\frac{2}{3}.$$

We know that $y > x = 3$. Using 2) and 3) as previously we deduced that $\frac{2}{3} = \frac{1}{y}+\frac{1}{z}+\frac{1}{t}+\frac{1}{u} < \frac{4}{y}$, i.e., $y < 6$. Thus $y \in \{4,5\}$. If we took $y = 5$, we would get $\frac{1}{z}+\frac{1}{t}+\frac{1}{u} = \frac{7}{15}$ and then $\frac{3}{z} > \frac{7}{15}$, i.e., $z < \frac{45}{7}$.

As $5 = y < z$ the only possible value for $z$ is 6. In this case

$$\frac{1}{t}+\frac{1}{u}=\frac{3}{10}$$

and, as $t < u$, it results that $\frac{2}{t} > \frac{3}{10}$, i.e., $t < \frac{20}{3}$. So $6 = z < t < \frac{20}{3}$, impossible, since $t$ is natural.

Therefore we cannot have $y = 5$ and so we can only have $y = 4$.

Going on in this manner we get $z = 5$, $t = 6$ and $u = 20$, which shows that

$$x = 3,\ y = 4,\ z = 5,\ t = 6,\ u = 20$$

is a solution for the problem.

This is actually the only solution of the problem, as it can be easily proven that we cannot have $x = 4$, and so on. □

## 21.3 Extending the First Source Problem

It is easy to notice the algorithmicity of the method used for solving Problem 21.1. It is clear that the same method can be applied to the much more general problem that follows.

**Problem 21.1a.** *Find the distinct natural numbers $n_1, n_2, \ldots, n_k$, satisfying the relation*

$$\frac{1}{n_1}+\frac{1}{n_2}+\ldots+\frac{1}{n_k}=1. \tag{1}$$

**Solution.** It is clear that we cannot solve this problem entirely by using only paper and a pen. For this reason, we describe the algorithm for solving this problem, which can be afterwards implemented in a programming language.

Because the numbers $n_1, n_2, \ldots, n_k$ are distinct, we can suppose that

$$n_1 < n_2 < \ldots < n_{k-1} < n_k.$$

Then

$$\frac{1}{n_1} > \frac{1}{n_2} > \ldots > \frac{1}{n_{k-1}} > \frac{1}{n_k} \tag{2}$$

and using (1) we get

$$\frac{k}{n_1} > 1, \text{ so } n_1 < k.$$

We cannot have $n = 1$, hence the set of possible values for $n_1$ is $M_1 = \{2, \ldots, k-1\}$. We choose a value for $n_1$ in this set, say $n_1 = a_1$. Substituting in (1) and doing some convenient changes, we get

$$\frac{1}{n_2} + \frac{1}{n_3} + \ldots + \frac{1}{n_k} > 1 - \frac{1}{a_1},$$

which together with (2) gives $n_2 < \frac{k-1}{b_1}$, where

$$b_1 = 1 - \frac{1}{a_1}. \tag{3}$$

Taking into account the fact that $n_1 < n_2$, for $n_1 = a_1$ (fixed), the set of possible values for $n_2$ will be

$$M_2 = \left\{a_1 + 1, \ldots, \left[\frac{k-1}{b_1}\right]\right\},$$

where $[x]$ is the integer part of $x$.

Following the same way, we get that, when starting with $n_1 = a_1$, the sets of possible values $M_1, M_2, \ldots, M_{k-1}$ for $n_1, n_2, \ldots, n_{k-1}$, respectively, are given by

$$M_i = \left\{a_{i-1} + 1, \ldots, \left[\frac{k-i+1}{b_{i-1}}\right]\right\}, \quad i = 1, 2, \ldots, k-1,$$

where

$$b_j = b_{j-2} - \frac{1}{b_{j-1}}, \quad j = 1, 2, \ldots, k-2,$$

$b_0 = 1$ and $b_1$ is given by (3). Finally we obtain the equation

$$\frac{1}{n_{k-1}} + \frac{1}{n_k} = b_{k-2}$$

in which we substitute $n_{k-1} = a_{k-1} \in M_{k-1}$ and we get

$$\frac{1}{n_k} = \frac{b_{k-2}a_{k-1} - 1}{a_{k-1}}, \text{ i.e., } \frac{1}{n_k} = \frac{m}{p}, \quad (m, p) = 1.$$

If $m = 1$, then taking $n_k = a_k = p \in \mathbb{N}$ we obtain a solution $(a_1, a_2, \ldots, a_k)$ of the equation (1). Further on, we proceed like this: we take, in turn, all the elements of the set $M_{k-1}$. When these are finished we take another value $a_{k-2} \in M_{k-2}$, that will determine a new set $M_{k-1}$ for which we evaluate the previous computations, and so on, until all the elements of the set $M_1$ are finished.

After having done this we get all the solutions of Problem 21.1a. □

**Remarks.** 1) Problem 21.1a can be reformulated as follows.

**Problem 21.1b.** *Having a half, a third, a quarter, a fifth,... of a given integer, make up this integer using $k(\geq 3)$ distinct fractions of it.*

2) The algorithm described above was implemented in 1981–1982 in **Fortran** on a **Felix C-256** computer (see also our paper [4]), and run for the values $k \in \{2,3,4,5,6,7\}$.

For $k = 7$, this variant of the program overran the memory limits, because of the high values of the last components of some of the solutions. Even for $k = 6$ we got fairly high numbers, for instance the last component of the solution is $n_6 = 3263442$;

3) Some quantitative aspects of the results obtained after running the program for the values $k \in \{3,4,5,6,7\}$ are gathered in Table 1.

| $k$ | Time | No. of | Remarks |
|---|---|---|---|
| | (sec) | sol. | |
| 5 | 3.5 | 1 | $n_1 \neq 2$. Sol. {3,4,5,6,20} |
| 5 | 15 | 72 | 71 sol. have $n_1 = 2$ and only one has $n_1 = 3$ |
| 4 | 15 | 6 | Two of the solutions: (2,3,7,42); (2,4,6,12) |
| 3 | 17 | 1 | Only one solution: (2,3,6). |
| 6 | 5 | 26 | $n_1 \neq 2$. Two of the solutions are |
| | | | (3,4,5,8,11,1320); (3,4,6,10,12,15) |
| 6 | 2 | 2317 | A "big" solution is |
| | 20 | | (2,3,7,43,1807, 3263442) |
| 7 | 180 | 945 | $n_1 \neq 2$. The "extreme" solutions are |
| | 10 | | (3,4,6,9,11,21,1772); (3,5,6,10,12,15,20) |

**Table 1**

A carefule analysis of these results for each value of $k$ reveals, surprisingly, that they exhibit some fundamental characteristics of theoretical mathematics, whether their presence is overt or only implicit.

**Problem 21.1c.** *If* $a_i \in \mathbb{N}^*$, $i = \overline{1,6}$ *are distinct and* $\sum_{i=1}^{6} \frac{1}{a_i} = 1$, *then* $\min\{a_1, a_2, \ldots, a_6\} \leq 3$.

**Problem 21.1d.** *If the polynomial* $P(x) = x^4 - ax^3 + bx^2 - cx + c$, *with* $a, b, c \in \mathbb{N}^*$, *has distinct natural roots, then*

*a)* $a \in [24, 54]$;

*b)* $P(x)$ *has the root* $x = 2$;

*c)* $c$ *is divisible by* 4.

**Problem 21.1e.** *The equation* $\frac{1}{x^2} + \frac{1}{y^2} + \frac{1}{z^2} + \frac{1}{t^2} + \frac{1}{u^2} = 1$ *does not have integer solutions.*

**Problem 21.1f.** If $a, b, c, d, e \in \mathbb{N}$ *are distinct natural numbers, whose product equals the sum of all the products of each four of them, then*

$$\min(\{a,b,c,d,e\} \setminus \min\{a,b,c,d,e\}) \leq 5.$$

**Problem 21.1g.** *Let* $n$ *be the product of the natural numbers* $n_1, n_2, \ldots, n_k$ *for which we have*

$$\frac{1}{n_1} + \frac{1}{n_2} + \ldots + \frac{1}{n_k} = 1 \quad (k \geq 3).$$

*Considering the factorization of* $n$,

$$n = p_1^{\alpha_1} p_2^{\alpha_2} \cdots p_m^{\alpha_m},$$

*prove that*

$$\alpha_1 + \alpha_2 + \ldots + \alpha_m \geq 2m.$$

**Remark.** We found elegant theoretical methods to solve all these problems — excepting for point a) of Problem 21.1d. Problem 21.1g generalizes a property observed in the cases $k \in \{3, 4, 5, 6, 7\}$.

We invite readers to try to implement the described algorithm in a programming language and then, analyzing the results, to discover new problems. We also recommend that they search for theoretical solutions to these new problems.

## 21.4 Another Type of Problem Obtained Using the Computer

**Problem 21.2.** *Find distinct digits* $a, b$ *and* $c$ *so that*

$$\overline{abc} \times \overline{abc} = \overline{nabc}, \tag{4}$$

*where* $n$ *is a digit or a natural number.*

**Solution.** From $c \times c = \ldots c$ we get $c \in \{0, 1, 5, 6\}$. If we took $c = 0$, then $\left(\overline{abc}\right)^2$ would end with $\ldots 00$, so we would have $b = c = 0$. But, as the digits are distinct, we can only have $c \in \{1, 5, 6\}$.

For $c = 1$, we get that the second last digit of the number $\left(\overline{abc}\right)^2$ is the last digit of $2b$ and must be equal to $b$. This is impossible.

Therefore $c \in \{5, 6\}$. Taking $c = 5$, we notice that the second last digit of the number $\left(\overline{abc}\right)^2$ is always 2 and so $b = 2$. Then, $\overline{a25} \times \overline{a25}$ has always the third last digit 6. Thus $a = 6, b = 2, c = 5$ is a solution for the problem and

$$625 \times 625 = 390625.$$

For $c = 6$ we follow the same steps and we get that $b$ should equal the last digit of the number $2b + 3$, which is verified only for $b = 7$. Evaluating $\overline{a76} \times \overline{a76}$

we deduce that $a$ should equal the last digit of the number $2a + 7$, which holds only for $a = 3$. Therefore, $a = 3$, $b = 7$ and $c = 6$ is the second solution for the problem and we have $376 \times 376 = 141376$, these two being the solutions of our problem. □

## 21.5 The Creative Approach

When I first solved this problem — more than 15 years ago — I immediately wondered if the relation (4) is still true for other combinations of the three digits in the right-hand sides, as for instance

$$\overline{nbac}, \overline{nbca}, \overline{ncba}, \overline{naaa}, \overline{nbbb}, \ldots. \tag{5}$$

Because the solution method of Problem 21.2 does not encourage anyone to investigate such a problem with pen on paper, and also because at that time I was working in a Computer Office as a programmer, I found that the most natural way of exploring this possibility was to use the computer.

So I designed a program in **Pascal**, that followed the same steps we take when multiplying "by hand" the numbers $\overline{abc}$ and $\overline{abc}$, and I made it cover all the possible cases $a, b, c \in \{0, 1, 2, \ldots, 9\}$ and $a \neq 0$, and verify each time whether the results were included in the list (5) or not.

Not only did I find new problems, but some of them were even more interesting than the original one, because they had a unique solution, while the initial problem had two. Here are some of these problems.

**Problem 21.2a.** *Find the digits $a, b, c$ for which there is a natural number $n$ so that*

$$\overline{abc} \times \overline{abc} = \overline{naaa}.$$

**Answer.** The only solution is $a = 4$, $b = 6$, $c = 2$ and we have $462 \times 462 = 213444$. □

**Problem 21.2b.** *Find the digits $a, b, c$ and the number $n$ so that*

$$\overline{abc} \times \overline{abc} = \overline{nbac}.$$

**Answer.** The problem has two solutions:

$$\begin{aligned} a &= 2, b = 0, c = 5 \text{ and } 205 \times 205 = 42025, \\ a &= 4, b = 2, c = 1 \text{ and } 421 \times 421 = 177241. \end{aligned}$$

**Problem 21.2c.** *The same question as before, for*

$$\overline{abc} \times \overline{abc} = \overline{ncba}.$$

**Answer.** The only solution is

$$a = 9, b = 6, c = 3 \text{ and } 963 \times 963 = 927369. \quad \square$$

Once we made the first step on our way to obtaining new problems of this type using the computer, it was very natural to explore new directions.

Below we give some examples of problems obtained in this way.

**Problem 21.2d.** *Find numbers* $\overline{ab}$ *divisible by* $9a + b$, *with* $a, b$ *decimal digits.*

**Answer.** The unique solution is $a = 7$ and $b = 9$. □

**Problem 21.2e.** *Find numbers* $\overline{ba}$ *for which* $\overline{ab}$ *is divisible by* $a + 8b$.

**Answer.** $a = 9, b = 8$. □

**Problem 21.2f.** *Find digits* $a, b, c, d$ *for which*

$$\overline{abcd} = \overline{cd} \times \overline{bcd}.$$

**Answer.** The unique solution is $a = 3, b = 1, c = 2, d = 5$. □

**Problem 21.2g.** *Reconstitute the multiplication* $\overline{abc} \times 3 = \overline{bcc}$.

**Answer.** The unique solution is $a = 1, b = 4, c = 8$. □

**Problem 21.2h.** *Find digits* $a, b, c$ *for which we have*

$$\overline{ac} \cdot \overline{b1} = \overline{1abc}.$$

**Solution.** The relation from the text can be rewritten as

$$c - 1 = \frac{100 + 9a}{b} - 10a \tag{6}$$

which shows that

$$\frac{100 + 9a}{b} \in \mathbb{N}. \tag{7}$$

We cannot have $c = 1$, since from (6) it would result that

$$100 = a(10b - 9)$$

and so $a = 4$ (because $10b - 9$ is odd) and $10b - 9 = 29$, which is impossible. Therefore, $c - 1 \geq 1$ and from (6) we deduce that

$$1 \leq \frac{100 + 9a}{b} - 10a \leq 8. \tag{8}$$

For $b \in \{7, 8\}$ it results from (8) that $a = 1$, for which (6) does not hold. For $b = 5$, from (7) it results that $a \in \{4, 8\}$, values which do not satisfy (8). Therefore, we are left with $b = 2$.

From (7) it results that $a$ is even, and from (8) that $11a \geq 84$, i.e., $a \geq 8$. Thus, $a = 8$ and then from (6) we get in the end $c = 7$, and so

$$a = 8, b = 2 \text{ and } c = 7$$

is the unique solution of the problem. □

**Remark.** We presented the solution of this problem in order to show that, even if these problems are formulated at the level of form V, solving them is not always an easy task. To challenge readers, we present now a list of problems of this kind in Table 2, in a short form and with hints or answers given, inviting them to solve all these by pencil on paper.

| Condition(s) | Solution(s) |
|---|---|
| $\overline{abcd} = \overline{ba} \cdot \overline{cd}$ | $1345 = 41 \cdot 35$ |
| $\overline{abc} = 5abc$ | $175 = 5 \cdot 1 \cdot 7 \cdot 5$ |
| $\overline{abc} : 7 = \overline{cbb}$ | $931 : 7 = 133$ |
| $\overline{abc} \cdot 3 = \overline{bbb}$ | $148 \cdot 3 = 444$ |
| $\overline{abc} \cdot 2 = \overline{bba}$ | $497 \cdot 2 = 994$ |
| $\overline{abc} \cdot 3 = \overline{bba}$ | $147 \cdot 3 = 441$ |
| $\overline{abc} \cdot 2 = \overline{bcc}$ | $250 \cdot 2 = 500$ |
| $\overline{abc} \cdot 3 = \overline{ccc}$ | $185 \cdot 3 = 555$ |
| $\overline{abc} : 4 = \overline{cbb}$ | $932 : 4 = 233$ |
| $\overline{ab} \cdot \overline{cd} = \overline{bcad}$ and | |
| $\overline{ad} \cdot \overline{cb} = \overline{abcd}$ | $a = 2, b = 1, c = 8, d = 7$ |
| $\overline{ab} \vdots (a + 6b), b \neq 0$ | $a = 5, b = 9$ |
| $\overline{abc} \times 6 = \overline{ccc}$ | $148 \times 6 = 888$ |
| $\overline{abcd} = (a + c) \cdot \overline{acd}$ | $a = 2, b = 4, c = 7, d = 5$ |
| $\overline{abc} = c \cdot \overline{ca}$ | $164 = 4 \cdot 41$ |
| $\overline{abc} = (a - b - c) \cdot \overline{cbb}$ | $a = 9, b = 3, c = 1$ |
| $\overline{abc} \vdots (a + b + c)^2$ and | |
| $\overline{cba} \vdots (a + b + c)^2$ | $a = 2, b = 4, c = 3$ |
| $\overline{abc} = 7abc$ | $a = 7, b = 3, c = 5$ |
| $\overline{abc} = 4abc$ | $a = 3, b = 8, c = 4$ |
| $\sqrt{\overline{abc}} = -a + b + c$ | $a = 1, b = 9, c = 6$ |
| $\sqrt{\overline{abc}} = \sqrt{\overline{ab}} + c$ | $a = 1, b = 6, c = 9$ |
| $\sqrt{a \cdot \frac{b}{c}} = a \cdot \sqrt{\frac{b}{c}}$ | $a = 2, b = 2, c = 3$ |
| | $a = 3, b = 3, c = 8$ |

**Table 2**

Another class of problems of the same type as the following can be obtained easily by means of a convenient program run on the computer.

**Problem 21.2i.** *Show that the second last digit of the number $6^n$ is odd for any $n \geq 2$.*

**Problem 21.2j.** *Prove that the second last digit of the number $47^n$ is even for any $n \in \mathbb{N}^*$.*

**Solution.** The last digit of the number $47^n$ might be 1, 7, 9, 3, as $n$ has one of the forms $4k, 4k+1, 4k+2, 4k+3,\ k \in \mathbb{N}$. Because the number $7 \cdot b$ has the second last digit even for any $b \in \{1,7,9,3\}$, it means that writing $47^n = \overline{\ldots ab}$, then from $47^{n+1} = \overline{\ldots ab} \times 47$ we get that the second last digit of the number $47^{n+1}$ is even, because $4 \cdot b + a$ is even, so its last digit is even as well.

As for $n = 1$ the number $47^1$ has the second last digit even, the claim results by induction. □

Other related problems are given in Table 3.

| Numbers | Second last digit | Third last digit |
|---|---|---|
| $56^n; 96^n; 86^n; 66^n$ | odd | – |
| $89^n; 87^n; 78^n; 67^n; 49^n$ | even | – |
| $76^n$ | 7 | odd |
| 26 | 7 | odd |

**Table 3**

## 21.6 Third Source Problem

The starting point of all explorations and investigations in this paragraph is

**Problem 21.3.** *Find out how much the product of four numbers changes if the first is increased with its half, the second is increased with its third, the third is decreased with its quarter and the fourth with its third.*

**Solution.** Let $a, b, c, d$ be the initial numbers. After modifying them as asked in the problem, these numbers will be, respectively

$$a + \frac{a}{2},\ b + \frac{b}{3},\ c - \frac{c}{4},\ d - \frac{d}{3},$$

i.e.,

$$a\left(1 + \frac{1}{2}\right),\ b\left(1 + \frac{1}{3}\right),\ c\left(1 - \frac{1}{4}\right),\ d\left(1 - \frac{1}{3}\right)$$

or

$$a \cdot \frac{3}{2},\ b \cdot \frac{4}{3},\ c \cdot \frac{3}{4},\ d \cdot \frac{2}{3}.$$

Their product is

$$a \cdot \frac{3}{2} \cdot b \cdot \frac{4}{3} \cdot c \cdot \frac{3}{4} \cdot d \cdot \frac{2}{3} = abcd,$$

a relation which shows that their product stays the same after these changes. □

**Remark.** The idea of solving this problem, though simple, is very beautiful. After the changes, one obtains fractions of the initial numbers, fractions whose product is 1:

$$\frac{3}{4} \cdot \frac{4}{3} \cdot \frac{3}{4} \cdot \frac{2}{3} = 1.$$

It is very natural then to explore how we could get $k$ fractions with the product 1. For each $k$ of the found ones we get a new problem. For $k = 3$, for instance, the fractions $\frac{6}{5}, \frac{5}{3}$ and $\frac{1}{2}$ verify the conditions asked for. We have thus obtained a new problem.

**Problem 21.3a.** *Let $a, b, c$ be three numbers. We increase the first one with its fifth, the second one with twice its third and decrease the third with its half. Show that the product of the obtained numbers equals that of the initial numbers.*

**Solution.** $\left(a+\frac{a}{5}\right)\cdot\left(b+\frac{2}{3}b\right)\cdot\left(c-\frac{1}{2}c\right) = a\cdot b\cdot c\cdot\frac{6}{5}\cdot\frac{5}{3}\cdot\frac{1}{2} = a\cdot b\cdot c.$ □

The easiest way of finding this type of fractions is to create a convenient algorithm, implement it in a programming language and run it on the computer. We worked in this manner some years ago, when we obtained lots of problems like this using a program written in **Fortran** and run on a **Felix C-256** computer.

In order to simplify things, we fixed from the start four successive numbers (although this condition was not necessary)

$$a, a+1, a+2, a+3$$

and we looked for the fractions $\frac{n_1}{n_{11}}, \frac{n_2}{n_{21}}, \frac{n_3}{n_{31}}, \frac{n_4}{n_{41}}$ so that

$$a(a+1)(a+2)(a+3)$$

$$= \left(a+\frac{n_1}{n_{11}}\right)\left(a+1+\frac{n_2}{n_{21}}\right)\left(a+2-\frac{n_3}{n_{31}}\right)\left(a+3-\frac{n_4}{n_{41}}\right).$$

In this manner we obtained the next two problems and many others (in Table 4 we give a few data with which one can formulate this kind of problems).

**Problem 21.3b.** *Show that*

$$123\cdot 124\cdot 125\cdot 126 = \left(123+\frac{3}{4}\right)\left(124+\frac{8}{33}\right)\left(125-\frac{3}{4}\right)\left(126-\frac{18}{71}\right).$$

**Problem 21.3c.** *Show that*

$$117\cdot 118\cdot 119\cdot 120 = \left(117+\frac{3}{5}\right)\left(118+\frac{11}{28}\right)\left(119-\frac{3}{5}\right)\left(120-\frac{15}{37}\right).$$

| $a$ | $n_1$ | $n_{11}$ | $n_2$ | $n_{21}$ | $n_3$ | $n_{31}$ | $n_4$ | $n_{41}$ |
|---|---|---|---|---|---|---|---|---|
| 117 | 1 | 4 | 50 | 67 | 1 | 4 | 72 | 95 |
| 117 | 1 | 9 | 55 | 62 | 1 | 9 | 96 | 107 |
| 333 | 3 | 4 | 22 | 89 | 3 | 4 | 48 | 191 |

**Table 4**

The path we have followed here is very restrictive, as we have taken $a, b, c, d$ successive numbers. If we do not take them like this, we have to find numbers $n_1, n_2, n_3, n_4, n_{11}, n_{21}, n_{31}, n_{41}$ for which

$$\left(1+\frac{n_1}{n_{11}}\right)\cdot\left(1+\frac{n_2}{n_{21}}\right)\cdot\left(1-\frac{n_3}{n_{31}}\right)\cdot\left(1-\frac{n_4}{n_{41}}\right)=1$$

and so we can get many more problems similar to the source problem.

## 21.7 Bibliographical Comments

Problem 21.1 was published as C:138 in *Gazeta Matematica* no. 10/1981, being taken from the Russian journal *Matematica v skole*, no. 3/1981.

Problem 21.1a is an extended form of Problem 21.1. It was the subject of some investigations materialized in the works [1], [2] and [3]. The last two of these were distinguished at those symposia. Paragraph 21.3 copies essentially our work [3]. Problem 21.1b was first published in [1].

Problem 21.1c was published in *Gazeta Matematica* no. 8/1984 as 20175, author V. Berinde.

Problem 21.2 was published in *Gazeta Matematica* no. 3/1987 as problem O.G: 44, authors being L. Pirsan and I. Gh. Tudor.

Problem 21.2a appeared in *Revista Matematica a Elevilor din Timisoara* no. 2/1987, as C7.7, and Problem 21.2b was published in *Gazeta Matematica* no. 2/1988 as OG:66, author V. Berinde. Problem 21.2d is problem 6221 from *Revista Matematica a Elevilor din Timisoara,* no. 2 / 1987, while Problem 21.2e appeared in *Gazeta Matematica* no. 3/1988, as problem OG: 68. Problem 21.2f is the problem E:9518* and Problem 21.2g is problem 70 for primary school, both published in *Gazeta Matematica,* no. 9/1988. Problem 21.2h was published as problem E:10404 in *Gazeta Matematica,* no. 1/1992.

Problem 9 in Table 1 was proposed at the Intercounty Mathematics Competition "Gr. Moisil", XV$^{th}$ Edition, Bistrita, 7–9 April 2000.

Problem 21.2i is problem 22253* from *Gazeta Matematica,* no. 1/1991. The author of all these problems is V. Berinde.

Problem 22.3 is the proposed problem VIII from a textbook for V$^{th}$ form in use in the period 1978–1996 [7], p. 248.

Problem 21.3b was published in *Gazeta Matematica,* no. 3/1989, as E: 9660*, author V. Berinde, while Problem 21.3c was proposed at the county round of the National Mathematics Competition, Maramures county, in the '90s.

## 21.8 References

1. **V. Berinde**, *Aspects of Probleming in the Interconnection Mathematics - Applied Mathematics.* (Romanian), The Symposium "Modern Aspects of Teaching some Mathematical Topics", Baia Mare, 21 December, 1983

2. **V. Berinde**, *Let Us Propose Mathematical Problems by Means of the Computer.* (Romanian), The County Session of Scientific Works, 5 November 1986, Baia Mare

3. **V. Berinde,** *Tracing New Exploring Directions by Means of the Computer.* (Romanian), The National Symposium of Young Scientists in Computer Science, Cluj-Napoca, November 1986

4. **V. Berinde**, *Let Us Propose Mathematical Problems by Means of the Computer.* (Romanian), Gazeta Matematica, **95** (1990), no. 8–9, 232–236

5. **V. Berinde**, *Let Us Discover Problems by Means of the Computer,* Mathematics&Informatics Quarterly, **9** (1999), no. 2, 70–71

6. **G.S. Lessels, M.I. Pelling**, *An Inequality for the Sum of Two Angle Bisectors and Median,* Publ. El. Fakulteta S. Mat i fiz., no. 577–598, 1977, 59–62

7. **C. Popovici et al.,** *Textbook for $V^{th}$ Form.* (Romanian), Editura Didactica si Pedagogica, 1987

8. **M. Cipu**, *A Problem for a Computer and Ceva,* Gazeta Matematica, **95** (2000), no. 9, 321–328

# Theme 22

# On the Convergence of Some Sequences of Real Numbers

## 22.1 Source Problem

It was a very difficult task to choose the source problem of this section, for many nice problems would deserve to occupy this position. The final choice was made because this kind of sequence is seldom published in the Romanian mathematical journals.

**Problem 22.** *Show that the sequence*

$$\sqrt{7}, \sqrt{7-\sqrt{7}}, \sqrt{7-\sqrt{7+\sqrt{7}}}, \sqrt{7-\sqrt{7+\sqrt{7-\sqrt{7}}}}, \ldots$$

*is convergent and find its limit.*

**Solution.** We take $u_1 = \sqrt{7}, u_2 = \sqrt{7-\sqrt{7}}$. The other terms of the sequence are given by the second order recurrence relation

$$u_{n+2} = \sqrt{7-\sqrt{7+u_n}}, \quad n \geq 1. \tag{1}$$

The sequence $(u_n)$ is well defined, that is, the quantity under the big square root in (1) is always positive. Indeed, we have

$$0 < u_n < \sqrt{7}, \text{ for all } n \in \mathbb{N}^*, \tag{2}$$

(which can be proved by induction) and then

$$\sqrt{7+u_n} < \sqrt{7+\sqrt{7}} < 7,$$

that is $7 - \sqrt{7+u_n} > 0$, $n \in \mathbb{N}^*$. Relation (2) also shows that the sequence $(u_n)$ is bounded.

Let us check the monotonicity of the sequence. We find that $u_1 > u_2 > u_3$ but $u_3 < u_4$, $u_4 > u_5$ and so on. If we compute $u_{2n+1} - u_{2n}$, by amplifying with the conjugates of the irrational expressions involved we get

$$signum\ (u_{2n+1} - u_{2n}) = -signum\ (u_{2n-1} - u_{2n-2}),\ n > 2,$$

a relation which shows that $(u_n)$ is not monotone. Therefore we cannot use the (BM) criterion, which states that a bounded and monotone sequence is convergent (see, for example, [5], Theorem II.5, p. 47). But we can successfully apply the "majorizing" criterion (M), given by Theorem II.4.1 ([5], p. 45):

**Theorem 1.** *Let $(a_n)$ be a sequence of real numbers. Suppose that there exist a real number $a$ and a sequence $(b_n)_{n\geq 0}$ of positive numbers convergent to zero, such that*

$$|a_n - b| \leq b_n,\ n \geq K\ (K \text{ fixed}). \tag{3}$$

*Then $(a_n)$ is convergent and*

$$\lim_{n\to\infty} a_n = a.$$

Indeed, to prove that $(u_n)$ is convergent, from (1) we derive

$$u_{n+2} - 2 = \sqrt{7 - \sqrt{7+u_n}} - 2 = \frac{3 - \sqrt{7+u_n}}{2 + \sqrt{7 - \sqrt{7+u_n}}}$$

or, after another similar transformation,

$$\begin{aligned} u_{n+2} - 2 &= \frac{2 - u_n}{\left(3 + \sqrt{7+u_n}\right)\left(2 + \sqrt{7 - \sqrt{7+u_n}}\right)} \\ &= \frac{2 - u_n}{\left(3 + \sqrt{7+u_n}\right)\left(2 + u_{n+2}\right)}. \end{aligned} \tag{4}$$

Now using (2) we deduce that

$$3 + \sqrt{7+u_n} \geq 3 \text{ and } 2 + u_{n+2} \geq 2,\ \forall n \in \mathbb{N}^*,$$

hence

$$\frac{1}{\left(3 + \sqrt{7+u_n}\right)\left(2 + u_{n+2}\right)} \leq \frac{1}{6},\quad \forall n \in \mathbb{N}^*.$$

Then, by (4) we have

$$|u_{n+2} - 2| \leq \frac{1}{6} \cdot |u_n - 2|,\ n \geq 1$$

from which we obtain by induction

$$|u_{2n} - 2| \leq \left(\frac{1}{6}\right)^{n-1} \cdot |u_2 - 2|,\ n \geq 1, \tag{5}$$

and

$$|u_{2n+1} - 2| \leq \left(\frac{1}{6}\right)^n \cdot |u_1 - 2|, \ n \geq 0, \tag{6}$$

respectively.

Now using Theorem 1, from (5) and (6) we obtain that the subsequences $(u_{2n})$ and $(u_{2n+1})$ both converge to 2. So, the sequence (3) is convergent and its limit is 2. □

## 22.2 A Close Examination of the Problem and its Solution

It is perhaps clear that the difficulty of Problem 22 is due, mainly, to the fact that the sequence $(u_n)$ is not monotone. If we succeed in proving, in a different manner, that the sequence $(u_n)$ is convergent, then by putting

$$x = \lim_{n\to\infty} u_n$$

and taking the limit in (1) we get that $x = 2$ is the unique solution of the equation

$$x = \sqrt{7 - \sqrt{7 + x}},$$

since in the left-hand side of the previous equation we have an increasing function, while in the right-hand side we have a decreasing function. This explains why above we considered the difference $u_{n+2} - 2$ in order to prove the convergence of $(u_n)$ in Problem 22.

At the beginning let us answer the following question: is the number 7 essential in Problem 20? In other words, could we replace it by an arbitrary number $a > 0$ such that the nature of the problem should not be affected?

In order to ensure that the square roots involved are well defined, it is needed that $a > \sqrt{a}$, which is true for $a > 1$, and furthermore we must have $\sqrt{a + \sqrt{a}} < a$, that is, $a + \sqrt{a} < a^2$. We take $\sqrt{a} = t$ and then we conclude that $t > 1$ and $t$ must also satisfy

$$f(t) \equiv t^4 - t^2 - 1 > 0.$$

As $f(1) = -1 < 0$ and $f(2) = 11 > 0$, it results that $f$ has a unique root $t_0$ on $(1, \infty)$, situated in the interval $(1, 2)$. Actually, as $f(\sqrt{2}) = 1 > 0$, it results that $t_0 < \sqrt{2}$. This shows that for any $t \geq \sqrt{2}$ we have $t^4 - t^2 - 1 > 0$ and hence, for any $a \geq 2$, we have $\sqrt{a + \sqrt{a}} < a$.

Therefore, following this direction of investigation, we get new problems of the same kind as Problem 22. Here are two of them.

**Problem 22a.** *Show that the sequence*

$$\sqrt{2}, \sqrt{2-\sqrt{2}}, \sqrt{2-\sqrt{2+\sqrt{2}}}, \sqrt{2-\sqrt{2+\sqrt{2-\sqrt{2}}}}, \ldots$$

*is convergent.*

**Solution.** In this case the recurrence relation is

$$u_{n+2} = \sqrt{2-\sqrt{2+u_n}}, n \geq 1, \text{ with } u_0 = \sqrt{2} \text{ and } u_1 = \sqrt{2-\sqrt{2}}.$$

The sequence is convergent but the limit of the sequence cannot be computed exactly, because the equation $x^2 = 2-\sqrt{2+x}$ has an irrational root in $(0, 1)$. □

This opens a new path for further investigations: find the values of $a$ such that, for the sequences defined by

$$u_{n-2} = \sqrt{a-\sqrt{a+u_n}},$$

the limit (if it exists) is a rational number. As the limit of such a sequence, when it exists, will be a solution of the equation

$$x = \sqrt{a-\sqrt{a+x}} \tag{7}$$

we are led to solve first the next problem.

**Problem 22b.** *Find $a \in \mathbb{N}^*$ for which the equation* (7) *has rational roots.*

**Solution.** We must have $x \geq 0$ and then (7) is equivalent to

$$\sqrt{a+x} = a - x^2. \tag{7'}$$

This equation has real roots only if $a - x^2 \geq 0$, that is $x \in [-\sqrt{a}, \sqrt{a}]$, and so we get that any solution of (7) must satisfy

$$x \in [0, \sqrt{a}]. \tag{8}$$

By squaring (7′) we obtain now

$$x^4 - 2ax^2 - x + a^2 - a = 0 \tag{7''}$$

which can also be regarded as a quadratic equation with respect to the unknown $a$:

$$a^2 - (2x^2+1)a + x^4 - x = 0,$$

whose roots are

$$a \in \{x^2 - 1,\ x^2 + x + 1\}.$$

In view of (8), we must have $x^2 \leq a$. It is clear that $x^2 \not< x^2 - 1$, but we have

$$x^2 < x^2 + x + 1.$$

This means that, if $a = m^2 + m + 1$, $m \in \mathbb{N}^*$, then equation (7) has the rational (actually integer) root

$$x = m.$$

□

We can restate this as

**Problem 22c.** *Prove that there exists an infinity of rational numbers* $a \in \mathbb{Q}_+^*$ *for which the equation*

$$x = \sqrt{x - \sqrt{a + x}}$$

*has a rational root.*

We are now in the position to go back to the problems of type 22a, in order to state them in a form closer to that of Problem 22.

**Problem 22d.** *Show that the sequence*

$$\sqrt{3}, \sqrt{3 - \sqrt{3}}, \sqrt{3 - \sqrt{3 + \sqrt{3}}}, \sqrt{3 - \sqrt{3 + \sqrt{3 - \sqrt{3}}}}, \ldots$$

*is convergent and find its limit.*

**Answer.** Here we have $m = 1$, $a = 3$ and, by the previous results, the limit of the sequence is 1. □

**Remark.** If we look carefully at the previous results, we notice that, starting with Problem 22, we have actually built a "problem factory"! Indeed, if we take a number $m \in \mathbb{Q}_+^*$ (not necessarily integer) and compute $a = m^2 + m + 1$, then

$$\sqrt{a}, \sqrt{a - \sqrt{a}}, \sqrt{a - \sqrt{a + \sqrt{a}}}, \sqrt{a - \sqrt{a + \sqrt{a - \sqrt{a}}}}, \ldots$$

will be a sequence of the same kind as the one in the source problem, for any $m \geq 1$ (which guarantees that $a \geq 2$).

Here are two brand new problems "produced" in this way.

**Problem 22e.** *Show that the sequence* $(u_n)_{n \geq 1}$,

$$u_{n+2} = \sqrt{\frac{19}{4} - \sqrt{\frac{19}{4} + u_n}}, \ n \geq 1 \quad \textit{and} \quad u_0 = \frac{19}{4}, \ u_1 = \sqrt{\frac{19}{4} - \sqrt{\frac{19}{4}}},$$

*converges to* $\frac{3}{2}$.

**Problem 22f.** *For any* $m \in \mathbb{Q} \cap (1, \infty)$, *the sequence* $(u_n)$,

$$u_0 = \sqrt{m^2 + m + 1}, \ u_1 = \sqrt{m^2 + m + 1 - \sqrt{m^2 + m + 1}}$$

*and*

$$u_{n+2} = \sqrt{m^2 + m + 1 - \sqrt{m^2 + m + 1 + u_n}}, \ n \geq 1,$$

*converges to* $m$.

At this stage, after discovering some new problems, we could be very satisfied and stop the investigations here. Would that be a good decision? Not at all! And here are some possible reasons. In the previous considerations we were concerned with superposed square roots of the same number, $a$. What happens with a sequence of one of the forms

$$\sqrt{b}, \sqrt{a-\sqrt{b}}, \sqrt{a-\sqrt{a+\sqrt{b}}}, \sqrt{a-\sqrt{a+\sqrt{a-\sqrt{b}}}}, \dots \tag{9}$$

or

$$\sqrt{b}, \sqrt{a-\sqrt{b}}, \sqrt{b-\sqrt{a+\sqrt{b}}}, \sqrt{a-\sqrt{b+\sqrt{a-\sqrt{b}}}}, \dots \tag{9'}$$

or

$$\sqrt{b}, \sqrt{a-\sqrt{b}}, \sqrt{b+\sqrt{a-\sqrt{b}}}, \sqrt{a-\sqrt{b+\sqrt{a-\sqrt{b}}}}, \dots ? \tag{9''}$$

What are the necessary assumptions on $a$ and $b$ to this end?

We begin with the simplest case.

**Problem 22g.** *Study the convergence of the sequence*

$$\sqrt{5}, \sqrt{3-\sqrt{5}}, \sqrt{3-\sqrt{3+\sqrt{5}}}, \sqrt{3-\sqrt{3+\sqrt{3-\sqrt{5}}}}, \dots .$$

**Solution.** We have in this case $u_0 = \sqrt{5}$, $u_1 = \sqrt{3-\sqrt{5}}$ and we notice that $u_{n+2} = \sqrt{3-\sqrt{3+u_n}}$, $n \geq 1$ is the same recurrence relation as in Problem 22d. Only the two initial terms are different. It is then clear that the limit of this sequence is also 1.

**Remark.** For a sequence of the form (9), where $a$ is a given number (say of the form $a = m^2 + m + 1$), we must have

$$\sqrt{b} \leq a \text{ and } \sqrt{b} \leq a^2 - a,$$

in order to guarantee its convergence. The limit will be, as previously, equal to $m$.

For the sequences of the form (9′) or (9″), a partial answer is given by the next problem.

**Problem 22g′.** *Show that the sequence* $(u_n)$, *given by* $u_1 = \sqrt{2}$ *and*

$$u_{n+1} = \begin{cases} \sqrt{6-u_n}, & \text{if } n \text{ is odd,} \\ \sqrt{u_n+2}, & \text{if } n \text{ is even} \end{cases}$$

*is convergent and find its limit.*

**Solution.** It is easy to show that $(u_n)$ satisfies the recurrence formulas

$$u_{2n+1} = \sqrt{2+\sqrt{6-u_{2n-1}}},\ n \geq 1,$$

and

$$u_{2n+2} = \sqrt{6-\sqrt{2+u_{2n}}},\ n \geq 1,$$

respectively. Using the same arguments as in solving Problem 22, we deduce that the subsequence $(u_{2n+1})$ is convergent and its limit is the solution of the equation

$$x = \sqrt{2+\sqrt{6-x}},$$

situated in the interval $\left[\sqrt{2}, 6\right]$, while the subsequence $(u_{2n})$ is convergent and its limit is the solution of the equation

$$y = \sqrt{6-\sqrt{2+y}},$$

situated in the interval $\left[-2, \sqrt{6}\right]$.

Since $x = 2$ and $y = 2$, respectively, are the unique roots of the equations above, we will obtain

$$\lim_{n\to\infty} u_{2n+1} = 2 = \lim_{n\to\infty} u_{2n}$$

which shows that $(u_n)$ is convergent in the whole and its limit is 2. □

**Remark.** In order to answer completely the question regarding the convergence of the sequences of the form (9′) or (9″), we shall try to find $a$ and $b$ such that the sequence $(u_n)$, $u_1 = \sqrt{b}$ and

$$u_{n+1} = \begin{cases} \sqrt{a-u_n}, & \text{if } n \text{ is odd,} \\ \sqrt{u_n+b}, & \text{if } n \text{ is even,} \end{cases}$$

has the same properties as the sequence in Problem 22g′.

Following the same steps as in solving Problem 22g′, we get that it is necessary that the equation

$$\sqrt{a-\sqrt{b+x}} = \sqrt{b+\sqrt{a-x}}$$

has a unique solution. By squaring the previous equation we obtain $a - b = \sqrt{a-x} + \sqrt{b+x}$ and by denoting $u = \sqrt{a-x}$ and $v = \sqrt{b+x}$ we get the system

$$\begin{cases} u+v = a-b, \\ u^2+v^2 = a+b, \end{cases} \Leftrightarrow \begin{cases} u+v = a-b, \\ uv = \dfrac{(a-b)^2-a-b}{2}. \end{cases}$$

Its solutions are obtained by solving the quadratic equation

$$t^2 - (a-b)t + \frac{(a-b)^2-a-b}{2} = 0.$$

In order to have a unique solution, we must have that

$$\Delta = (a-b)^2 - 4\left[\frac{(a-b)^2 - a - b}{2}\right] = 2\left[2a + 2b - (a-b)^2\right]$$

is zero, and thus we are led to the equation ($a$ is the unknown)

$$a^2 - 2(b+1)a + b^2 - 2b = 0$$

which has integer roots only if $\Delta' = 4b+1$ is a perfect square, that is $4b+1 = k^2 \Leftrightarrow b = \dfrac{k^2-1}{4}$, which leads to $k = 2p+1, k \in \mathbb{N}$. Then we obtain $b = p(p+1)$ and $a = p^2 + 3p + 2$ or $a = p^2 - p$.

As $u + v > 0$, we must have $a > b$, and thus the values for $a$ and $b$ are

$$\begin{cases} a = p^2 + 3p + 2 = (p+1)(p+2), \\ b = p^2 + p = p(p+1), \end{cases} \quad p \in \mathbb{N},$$

and the unique solution of the initial equation will be $x = p+1$ (the limit of the sequence). In this way we obtained a new "problem factory" which produces, for $p = 1$, exactly Problem 22g′. It is contained in

**Problem 22g″.** *For any $p \in \mathbb{N}^*$, the sequence $(u_n)$,*

$$u_1 = \sqrt{p(p+1)}, \quad u_{n+1} = \begin{cases} \sqrt{(p+1)(p+2) - u_n}, & \text{if } n \text{ odd}, \\ \sqrt{u_n + p(p+1)}, & \text{if } n \text{ even}, \end{cases}$$

*$n \geq 1$, converges to $p+1$.*

At the end of this paragraph, we suggest that readers study the properties (limits, continuity etc.) of the function $f : [2, \infty) \to \mathbb{R}$, defined by

$$f(a) = \lim_{n\to\infty} u_n(a),$$

where $(u_n(a))$ is the sequence

$$\sqrt{a}, \sqrt{a - \sqrt{a}}, \sqrt{a - \sqrt{a + \sqrt{a}}}, \sqrt{a - \sqrt{a + \sqrt{a - \sqrt{a}}}}, \ldots.$$

We know so far that $f(3) = 1$, $f(7) = 2$, $f\left(\dfrac{19}{4}\right) = \dfrac{3}{2}$ and, in general, $f(m^2 + m + 1) = m$, for any $m \in \mathbb{Q}$, $m \geq 1$.

## 22.3 A Necessary and Sufficient Criterion for Convergence

Let us notice some collateral facts related not primarily to the sequences we considered in the previous paragraph, but to the method we used to solve the problems included there.

The method is essentially based on the $(M)$ criterion for sequences ($M$ stands for majorizing), stated in Theorem 1. The $(M)$ criterion is included in [5], in the subparagraph "Sufficient Criteria for the Convergence of Sequences", alongside of the $(BM)$ criterion. While the $(BM)$ criterion is indeed a sufficient criterion (for example, the sequence $(a_n)_{n\geq 1}, a_n = (-1)^n \cdot \dfrac{1}{n}, n \in \mathbb{N}^*$ is convergent but is not monotone), the $(M)$ criterion is actually a **necessary and sufficient** criterion, as shown by

**Theorem 2.** *A sequence of real numbers $(a_n)_{n\geq 1}$ is convergent to $a$ if and only if there exists a sequence of positive numbers $(b_n)_{n\geq 1}$, converging to zero, such that* (3) *holds.*

**Proof.** The "sufficiency" part follows by Theorem 1. In order to prove the necessity we take the sequence

$$b_n = |\, a_n - a\,|\,,\ n \in \mathbb{N},$$

which obviously satisfies (3). □

**Remark.** The theoretical superiority of the $(M)$ criterion with respect to the $(BM)$ criterion is revealed only in those circumstances when the sequence $(a_n)$ is not monotone — so we cannot apply the $(BM)$ criterion — and, moreover, we are able to construct in a nontrivial way the sequence $(b_n)$, involved in (3).

There are several situations — as we have shown in Section 13, Solution 3 of Problem 13 — when we can still apply the $(BM)$ criterion, but separately, to the odd and even subsequences, respectively, which are both monotonic. This does not happen in the case of Problem 22, because $(u_{2n})$ and $(u_{2n+1})$ are not monotonic. For the following sequence

$$a_n = \frac{1}{n}\sin n,\ \ n \geq 1 \tag{10}$$

we can use only the $(M)$ criterion, since

$$\left|\frac{1}{n}\sin n - 0\right| \leq \frac{1}{n},\ \ n \geq 1.$$

Neither $(a_{2n})$ nor $(a_{2n+1})$ are monotonic, so we cannot apply the $(BM)$ criterion. We shall pay more attention to the sequences of the form (10) in the next paragraph. Now let us come back to the **necessary and sufficient** criterion $(M)$.

It is clear that it can be formulated for a complex sequence $(a_n)$ as well, and that its correspondent in the case of sequences of functions is the well known Weierstrass uniform convergence test. This one, a very useful test in applications, is in fact, as shown in the $(M)$ criterion, a necessary and sufficient test for uniform convergence even if in practice we use only its sufficiency part. Thus we obtained the following improved reformulations of the Weierstrass test (Criterion II and its Corollary [7], p. 722 — for the sequences of functions, and Criterion II, p. 741 and its Corollary, p. 742 in [7] — for the series of functions, respectively).

We state here only two of these criteria.

**Theorem 3.** *Let* $(f_n)$ *be a sequence of functions,* $f_n : A \to \mathbb{R}$, $n \in \mathbb{N}$ *and* $f : A \to \mathbb{R}$. *Then* $f_n \rightrightarrows f$ *if and only if there exists a sequence of positive functions* $(\varphi_n), \varphi_n : A \to \mathbb{R}$, *uniformly convergent to zero, such that*

$$|f_n(x) - f(x)| \leq \varphi_n(x), \quad \forall\, n \geq K \ (K \text{ fixed}), \ \forall x \in A.$$

**Theorem 4.** *Let* $\sum f_n$ *be a series of functions defined on* $A$. *The series is uniformly convergent on* $A$ *if and only if there exists a convergent series of positive terms* $\sum a_n$ *such that*

$$|f_n(x)| \leq a_n, \ \forall \ n \geq K \ (K \text{ fixed}), \ \forall x \in A.$$

We invite readers to formulate by themselves the other two improved variants of the Weierstrass test.

We notice that in the form given by Theorems 3 and 4, the Weierstrass test offers a characterization of the uniformly convergent sequences and series of functions, respectively.

## 22.4 The Rate of Convergence of Sequences of Real Numbers

When solving Problem 22 and its relatives we used the $M$ criterion in the formulation given by Theorem 1. Condition (3) in this theorem illustrates a certain relationship between the convergence of the majoring sequence $(b_n)$ to zero, on the one hand, and the convergence of the sequence $(a_n)$ to $a$, on the other hand. We shall try to describe this fact in a suitable framework, that includes the concept of **rate of convergence**.

Intuitively, among the sequences $\left(\frac{1}{n}\right)_{n\geq 1}$ and $\left(\frac{1}{2^n}\right)_{n\geq 1}$, both converging to zero, the second one converges faster, in the sense that its terms approach quicker to zero than the terms of the first sequence. We are thus led to introducing the following concept.

**Definition 1.** *Let* $(a_n), (b_n)$ *be two sequences of positive numbers, both convergent to zero, such that the limit*

$$\lim_{n\to\infty} \frac{a_n}{b_n} = l \tag{11}$$

*exists. Then*

*a) if* $l = 0$, *we shall say that* $(a_n)$ *converges* **faster** *than* $(b_n)$;

*b) if* $l \neq 0$ *and* $l$ *is finite, we shall say that* $(a_n)$ *and* $(b_n)$ *have the same convergence rate.*

**Remarks.** 1) If $l = \infty$, that is $\lim\limits_{n\to\infty} \frac{a_n}{b_n} = 0$, then in view of a) the sequence $(b_n)$ converges faster than $(a_n)$. This is the case of the two sequences we considered above, since

$$\lim_{n\to\infty} \frac{1}{n} / \frac{1}{2^n} = \lim_{n\to\infty} \frac{2^n}{n} = \infty.$$

2) The concept of convergence rate given by Definition 1 in the family of null sequences (i.e., sequences converging to zero) can be extended for any pair of convergent sequences. Indeed, if $(u_n)$ and $(v_n)$ are two sequences converging to $u$ and $v$, respectively, then the sequences $(a_n), (b_n)$, given by $a_n = |u_n - u|, b_n = |v_n - v|, n \in \mathbb{N}$, are both null sequences. Assuming that the limit

$$\lim_{n\to\infty} \frac{|u_n - u|}{|v_n - v|} = l \tag{11'}$$

exists, we have the following three possible situations:

a) $l = 0$, when $(u_n)$ converges faster (to $u$) than the sequence $(v_n)$ (to $v$);

b) $l \neq 0$ and $l \neq \infty$, when $(u_n)$ and $(v_n)$ have the same convergence rate;

c) $l = \infty$, when $(v_n)$ converges faster than $(u_n)$.

3) For the sequences $(a_n), (b_n)$, converging to infinity, the concept of convergence rate may be defined similarly, supposing that the limit (11) does exist. In fact, it is easy to see that the condition that these sequences are positive, in Definition 1, as well as the absolute value in (11′) can be removed.

**Examples.** 1) In view of our concept, the relation

$$\lim_{n\to\infty} n(\ln 2 - a_n) = \frac{1}{4} \Leftrightarrow \lim_{n\to\infty} \frac{|a_n - \ln 2|}{\frac{1}{n}} = \frac{1}{4} \tag{12}$$

shows that the sequence $(a_n)$,

$$a_n = \frac{1}{n+1} + \frac{1}{n+2} + \ldots + \frac{1}{2n}, \quad n \geq 1,$$

which converges to $\ln 2$, has the same convergence rate as the sequence $\left(\frac{1}{n}\right)_{n\geq 0}$.

2) Let $(x_n)_{n\geq 1}$ be the sequence

$$x_{n+1} = \frac{1}{2}\left(x_n + \frac{2}{x_n}\right), \quad n \geq 1, \ x_1 > 0,$$

which represents Newton's iterations for solving the quadratic equation $x^2 - 2 = 0$ in $(0, \infty)$, i.e., for approximating the value of $\sqrt{2}$. This sequence is decreasing and $x_n \to \sqrt{2}$, as $n \to \infty$ (see, for example, [5], pp. 67–68), and so $x_n \geq \sqrt{2}$, for all $n \geq 2$. By induction we prove that

$$x_{n+1} - \sqrt{2} \leq \frac{1}{2}\left(x_n - \sqrt{2}\right), \quad n \geq 2,$$

from which we obtain

$$0 \leq x_n - \sqrt{2} \leq \frac{1}{2^n}\left(x_1 - \sqrt{2}\right), \ n \geq 2.$$

This shows that

$$0 \leq n^k\left(x_n - \sqrt{2}\right) \leq \frac{n^k}{2^n}\left(x_1 - \sqrt{2}\right), \quad n \geq 2, \ k \in \mathbb{N},$$

and hence, since $\lim\limits_{n\to\infty} \frac{n^k}{2^n} = 0, k \in \mathbb{N}$ ($k$ fixed), we get

$$\lim_{n\to\infty} n^k\left(x_n - \sqrt{2}\right) = 0, \quad (k \text{ fixed})$$

and this last relation expresses a very important thing from the point of view of approximation theory:

The sequence $(x_n)$ converges **extremely fast** to $\sqrt{2}$, even faster than any sequence $\left(\frac{1}{n^k}\right)$ to zero, $k \in \mathbb{N}$.

For this reason, starting from any initial guess $x_1 > 0$ of $\sqrt{2}$, the iterate $x_n$ will be a good approximation of $\sqrt{2}$, for small values of $n$. Indeed, starting from $x_1 = 1$, after three iterations we obtain the value of $\sqrt{2}$ with 3 exact digits

$$x_3 = \frac{577}{408} \approx 1.4142.$$

Unlike the previous sequence, the following one given by $a_1 = 1$ and

$$a_{n+1} = a_n + (-1)^n \cdot \frac{1}{n+1}, \ n \geq 1$$

converges **very slowly** to $\ln 2$, in view of (12), which shows that this sequence has the same convergence rate as $\left(\frac{1}{n}\right)$.

Indeed, the first term of $(a_n)$ which will approximate $\ln 2$ with two exact digits will be $a_{999}$!

Now we go back to Problem 22, to restate it in a more complete form as

**Problem 22h.** *Let* $(u_n)$ *be the sequence defined by* $u_1 = \sqrt{7}, u_2 = \sqrt{7-\sqrt{7}}$ and $u_{n+2} = \sqrt{7 - \sqrt{7 + u_n}}$, $n \geq 1$. *Show that*

$$\lim_{n\to\infty} n^k(u_n - 2) = 0, \quad \forall k \in \mathbb{N}. \tag{13}$$

**Solution.** From the solution of Problem 22, the relations (5) and (6), we get

$$n^k \cdot |u_{2n} - 2| \leq \frac{n^k}{6^{n-1}} \cdot |u_2 - 2|, \quad n \geq 1$$

and

$$n^k \cdot |u_{2n+1} - 2| \leq \frac{n^k}{6^n} \cdot |u_1 - 2|, \quad n \geq 0,$$

respectively, which proves (13), since $\lim\limits_{n\to\infty} \frac{n^k}{6^n} = 0$, $\forall k \in \mathbb{N}$ (fixed). □

**Remarks.** 1) The obtained relation (13) shows that $(u_n)$ converges to 2 **faster** then any sequence of the form $\left(\frac{1}{n^k}\right)$ to zero, $k \in \mathbb{N}^*$. Similar results can be obtained for the sequences of the form 22g′.

2) It is important to stress the fact that not just any two sequences could be compared by means of the concept of convergence rate given by Definition 1. For example, the sequences $(a_n)$, $a_n = \frac{1}{n}\sin n$, $n \geq 1$ and $(b_n)$, $b_n = \frac{1}{n}$, $n \geq 1$, for which we have $a_n \to 0$, $b_n \to 0$ (as $n \to \infty$) cannot be compared, since the limit

$$\lim_{n\to\infty} \frac{a_n}{b_n} = \lim_{n\to\infty} \sin n \tag{14}$$

does not exist.

This fact illustrates the significant difference between the behaviour of the two sequences when approaching their common limit, zero.

In fact, for any sequence whose convergence was established by the $M$ criterion, we also get indirect information on its convergence rate. Based on the results in Section 22.3 we also rediscover the concept of convergence rate for sequences and series of functions, a concept that is well known in the mathematical literature (see [6], [8]).

Let us go back to a previous statement, i.e., that the limit (14) does not exist. We assume that readers know the definition of the limit of a function by means of sequences (Heine): the limit

$$\lim_{n\to\infty} f(x) = l \tag{15}$$

exists if and only if for any sequence $(x_n)$, $x_n \to a$ as $n \to \infty$, we have

$$\lim_{n\to\infty} f(x_n) = l. \tag{15′}$$

If there exist two sequences, $(x'_n)$ and $(x''_n)$, both converging to $a$, such that

$$\lim_{n\to\infty} f(x'_n) = l', \quad \lim_{n\to\infty} f(x''_n) = l''$$

and $l' \neq l''$, then we conclude that the limit

$$\lim_{n\to\infty} f(x)$$

does not exist.

This is the way in which, for instance, we prove that $\lim\limits_{n\to\infty} \sin\frac{1}{x}$ does not exist (see also Theme 18, Remark 1) at the end of Section 18.1). Indeed, taking $x'_n = \frac{1}{2n\pi}$, $x''_n = \frac{2}{(2n+1)\pi}$, $n \in \mathbb{N}^*$, we get $\lim\limits_{n\to\infty} \cos\frac{1}{x'_n} = 1 \neq 0 = \lim\limits_{n\to\infty} \cos\frac{1}{x''_n}$.

A typical situation when the implication (15) $\Rightarrow$ (15′) is used in practice appears in the case when we apply l'Hospital's rule to compute various limits of sequences, a procedure which is generally applied in a wrong way by the students. Let us consider, for example, the limit

$$\lim_{n\to\infty} \frac{\ln(a+n)}{n}, \quad a > 0.$$

We cannot use **directly** l'Hospital's rule by writing

$$\lim_{n\to\infty} \frac{\ln(a+n)}{n} = \lim_{n\to\infty} \frac{\frac{1}{a+n}}{1} = \lim_{n\to\infty} \frac{1}{a+n} = 0,$$

since $\mathbb{N}$ is a discrete set (the derivative is not defined). The correct procedure is the following one: we compute by l'Hospital's rule the limit in $\mathbb{R}$

$$\lim_{x\to\infty} \frac{\ln(a+x)}{x} = \lim_{x\to\infty} \frac{1}{a+x} = 0,$$

and hence, for any sequence $(x_n)$, $x_n \to \infty$ (as $n \to \infty$), we have

$$\lim_{n\to\infty} \frac{\ln(a+x_n)}{x_n} = 0,$$

which in particular, for $x_n = n, n \geq 1$, gives

$$\lim_{n\to\infty} \frac{\ln(a+n)}{n} = 0.$$

Coming back now to the sequence $(\sin n)_{n\geq 1}$, the following problem arises: knowing that

$$\lim_{x\to\infty} f(x) \tag{16}$$

does not exist, is it then true that

$$\lim_{n\to\infty} f(n) \tag{16'}$$

does not exist either?

Unlike the case of the existence of the limit when the equivalence (15) $\Leftrightarrow$ (15′) holds, in the case of the nonexistence of the limit (16) **it does not result** that the limit (16′) does not exist either, as shown by the next example.

The function $f : \mathbb{R} \to \mathbb{R}, f(x) = \{x\}$ (the fractional part of $x \in \mathbb{R}$), does not have a limit at $+\infty$, being periodic, but

$$\lim_{n\to\infty} f(n) = 0$$

since $f(n) = 0$, $\forall n \in \mathbb{N}^*$.

Therefore, even if we know that for $f(x) = \sin \frac{1}{x}$ the limit (16) does not exist, we cannot conclude from this that the limit (16′) does not exist either. We have to prove it directly, that is to solve the next problem.

**Problem 22i.** *Show that the sequence* $(\sin n)_{n\geq 1}$ *has no limit.*

**Solution.** Let

$$A = \left\{ n \in \mathbb{N} | \exists\, k \in \mathbb{N}, \left(2k + \frac{1}{6}\right)\pi < n < \left(2k + \frac{5}{6}\right)\pi \right\}$$

and

$$B = \left\{ n \in \mathbb{N} | \exists\, k \in \mathbb{N}, \left(2k + \frac{7}{6}\right)\pi < n < \left(2k + \frac{11}{6}\right)\pi \right\}.$$

We show first that both sets $A$ and $B$ are infinite. Indeed, for any positive integer $k \in \mathbb{N}$, the integer

$$n = \left[\left(2k + \frac{5}{6}\right)\pi\right], \quad [x] = \text{ the integer part of } x,$$

belongs to $A$, since

$$n < \left(2k + \frac{5}{6}\right)\pi < n + 1 \text{ means } n > \left(2k + \frac{5}{6}\right)\pi - 1$$

and so from $\left(2k + \frac{5}{6}\right)\pi - 1 > \left(2k + \frac{1}{6}\right)\pi$ (which reduces to $2\pi > 3$), we get

$$\left(2k + \frac{1}{6}\right)\pi < n < \left(2k + \frac{5}{6}\right)\pi$$

and therefore $A$ is infinite. Similarly, letting

$$n = \left[\left(2k + \frac{11}{6}\right)\pi\right],$$

we prove that $B$ is infinite as well.

But all the terms of the subsequence $(\sin n)_{n\in A}$ belong to the interval $\left(\frac{1}{2}, 1\right]$, while all the terms of the subsequence $(\sin n)_{n\in B}$ belong to the interval $\left[-1, -\frac{1}{2}\right)$. This shows that the sequence $(\sin n)_{n\geq 1}$ cannot be convergent. □

**Problem 22j.** *Study the convergence of the sequence* $(n^a \cos n)_{n\in\mathbb{N}}$, $a \in \mathbb{R}$.

**Solution.** 1) For $a < 0$ we have $\lim_{n\to\infty} n^a \cos n = \lim_{x\to\infty} x^a \cos x = 0$, since $|x^a \cos x| \leq |x|^a$.

2) If $a \geq 0$, then the function $f(x) = x^a \cos x$ does not have a limit at $+\infty$. We prove that neither does the sequence. To this end we consider the sets of positive integers

$$A = \left\{ n \in \mathbb{N} | \exists\, k \in \mathbb{N}, \left(2k - \frac{1}{3}\right)\pi < n < \left(2k + \frac{1}{3}\right)\pi \right\}$$

and

$$B = \left\{ n \in \mathbb{N} | \exists\, k \in \mathbb{N}, \left(2k + \frac{2}{3}\right)\pi < n < \left(2k + \frac{4}{3}\right)\pi \right\}.$$

We prove, as previously, that $A$ and $B$ are infinite sets and then we consider two different cases.

2a) If $a > 0$, then the subsequence $(n^a \cos n)_{n \in A}$ converges to $+\infty$, since from $\cos n > \frac{1}{2}$, for $n \in A$, we get

$$n^a \cos n > \frac{n^a}{2} \to \infty,$$

while the subsequence $(n^a \cos n)_{n \in B}$ converges to $-\infty$, in view of the fact that $n \in B$ implies $\cos n < -\frac{1}{2}$, that is $n^a \cos n < -\frac{n^a}{2}$.

2b) If $a = 0$, the sequence reduces to $(\cos n)_{n \geq 0}$. In this case, the interval $\left[-1, -\frac{1}{2}\right)$ contains all the terms of the subsequence $(\cos n)_{n \in A}$, while the interval $\left(\frac{1}{2}, 1\right]$ contains all the terms of the subsequence $(\cos n)_{n \in B}$. This shows that $(\cos n)$ cannot have a limit (otherwise, any neighborhood of this limit would contain all the terms of the sequence, except for a finite number of them). □

At the end of this section we invite readers to continue their investigations, in order to discover themselves other related problems, and to solve the following ones.

**Problem 22k.** *Study the convergence of the sequence* $(n^a \sin n)_{n \in \mathbb{N}}$, $a \in \mathbb{R}$.

**Problem 22l.** *Prove that the sequence* $(a_n)_{n \geq 1}$, $a_n = \cos(\ln n)$ *is not convergent.*

**Problem 22m.** *Study the convergence of the sequence*

$$(n^a \cos \sqrt{n})_{n \geq 1}, \quad a \in \mathbb{R}.$$

## 22.5 Bibliographical Comments

Problem 22 was proposed at the Putnam Mathematics Competition in the U.S.A., in 1953. It is taken from [4]. The content of Section 22.1 is taken from our paper [2], with some minor modifications.

Problem 22g was proposed by the author at the final round of the 51st National Mathematical Olympiad, Brasov 21–28 April 2000, and was discussed by the jury.

The first part of Section 22.3 is taken from our paper [2], where the necessary and sufficient aspect of the majorizing criterion was presented. Theorems 3 and 4 appear in this book for the first time, being natural consequences of Theorem 2 in [2].

The content of Section 22.4 is an ad-hoc variant of our paper [3].

Problem 22j was published by the author in *Gazeta Matematica* no. 3/1991 as Problem C:1108, while Problem 22l appeared in *Gazeta Matematica* no. 4/2000 as Problem 24307, its author being C. Mortici.

## 22.6 References


1. **V. Berinde,** *On a Competition Problem.* (Romanian), Gazeta Matematica, **90** (1985), no. 6, 186–188

2. **V. Berinde,** *On a Necessary and Sufficient Criterion for the Convergence of Sequences.* (Romanian), ASTRA Matematica, **1** (1990), no. 5, 18–21

3. **V. Berinde,** *On the Convergence Rate of Sequences of Real Numbers.* (Romanian), Gazeta Matematica, **103** (1998), no. 4, 146–153

4. **A.M. Gleason,** *The W.L. Putnam Mathematical Competition.* The Mathematical Association of America, 1980

5. **Gh. Gussi, O. Stanasila, T. Stoica,** *Elements of Mathematical Analysis. Textbook for XI Form.* (Romanian), Editura Didactica si Pedagogica, 1991

6. **K. Knopp,** *Theorie und Anwendung der Unendlichen Reichen,* Springer Verlag, 1964

7. **M. Nicolescu, N. Dinculeanu, S. Marcus,** *Mathematical Analysis.* (Romanian), $4^{th}$ Edition, Editura Didactica si Pedagogica, 1972

8. **N.N. Vorobiev,** *Theory of Series.* (Russian), Izdatelstvo "Nauka", Moskova, 1975

9. **S. Cremarenko, B. Enescu,** *Using the Computer in Studying the Convergence of Some Recurrent Sequences.* (Romanian), Gazeta Matematica, **93** (1988), no. 11–12, 434–437

10. **G.E. Shilov,***Elementary Real and Complex Analysis.* Dover Publications, New York, 1996

# Theme 23

# An Application of the Integral Mean

The aim of this section is to introduce the notion of *integral mean at a point* as well as to illustrate some of its properties and then to apply the working technique given by this notion to solving a problem, in order to highlight its creative virtues. In this way we will get some more general results related to a class of functional equations.

## 23.1 Source Problem

**Problem 23.1.** *Let $f : [0,\infty) \to [0,\infty)$ be an increasing function with the property that there exists $a \in (0,1)$ so that*

$$\int_0^x f(t)\,dt = \int_0^{ax} f(t)\,dt, \quad \forall\ x \in [0,\infty). \tag{1}$$

*Prove that $f(x) = 0$, for any $x \in [0,\infty)$.*

**Solution 1.** Being monotonic, $f$ can be integrated on any interval $[0,x]$, $x > 0$ (that is, it is locally integrable). Changing $u(t) = \dfrac{t}{a}$ in the integral on the right-hand side of relation (1) we get

$$\int_0^{ax} f(t)\,dt = a\int_0^{x} f(at)\,dt$$

and so (1) becomes

$$\int_0^x [f(t) - af(at)]\,dt = 0, \ \ \forall t \in [0,\infty).$$

Because $a \in (0,1)$, it results that $at < t$, $\forall t \in [0,x]$ and using the fact that $f$ is increasing we deduce that $f(at) \leq f(t) \Leftrightarrow -f(at) \geq -f(t)$ and so

$$f(t) - af(at) \geq (1-a)f(t), \ \ \forall t \in [0,x].$$

We use now the monotonicity of the integral with respect to the integrand to obtain

$$0 = \int_0^x [f(t) - af(at)]\,dt \geq (1-a)\int_0^x f(t)\,dt. \tag{2}$$

Let $t_0 \in [0,x)$. Then, as $f(t) \geq 0$, $\forall t \in [0,\infty)$, we have

$$\int_0^x f(t)\,dt = \int_0^{t_0} f(t)\,dt + \int_{t_0}^x f(t)\,dt \geq \int_{t_0}^x f(t)\,dt. \tag{3}$$

But $f$ is increasing and so

$$f(t) \geq f(t_0), \ \forall t \in [t_0, x]$$

from which we deduce that

$$\int_{t_0}^x f(t)\,dt \geq f(t_0)\cdot(x - t_0)$$

and from (2) and (3) we have

$$(1-a)\cdot(x-t_0)\cdot f(t_0) \leq 0.$$

Because $1-a > 0$, $t - x_0 > 0$ and $f(t_0) \geq 0$, it results that $f(t_0) = 0$ and as $t_0$ was arbitrarily taken, we conclude that $f(x) = 0$, $\forall\, x \in [0,\infty)$. □

**Solution 2.** The fact that $f$ is integrable ensures that

$$F(x) = \int_0^x f(t)\,dt, \ \ x \in [0,\infty)$$

is a continuous function [4] and, moreover, $F(0) = 0$. Relation (1) can be rewritten as

$$F(x) = F(ax)$$

from which we get that

$$F(x) = F(ax) = F(a^2x) = \ldots = F(a^n x), \ \forall n \in \mathbb{N}, \ \forall\, x \in [0, \infty)$$

and so

$$F(x) = \lim_{n\to\infty} F(a^n x).$$

But $a^n x \to 0$, when $n \to \infty$, $\forall x \in [0, \infty)$, and $F$ is continuous at zero, which means that

$$\lim_{n\to\infty} F(a^n x) = F(0), \text{ and so } F(x) = 0, \ \forall x \in [0, \infty).$$

Therefore

$$\int_0^x f(t)\,dt = 0, \ \forall x \in [0, \infty).$$

We take $t_0 \in [0, x)$ and then, as $f$ is increasing, we obtain

$$0 = \int_0^x f(t)\,dt = \int_0^{t_0} f(t)\,dt + \int_{t_0}^x f(t)\,dt \geq \int_{t_0}^x f(t)\,dt \geq (x - t_0) \cdot f(t_0)$$

from which we deduce that $f(t_0) \leq 0$, i.e., $f(t_0) = 0$. Therefore $f \equiv 0$. □

**Remark.** The first question we can now ask is whether the conclusion of Problem 23 stays the same in case $f$ is decreasing. We cannot give an answer using Solution 1, but Solution 2 can be adapted as follows.

The integrability of $f$ (and so the continuity of $F$) stays the same in case $f$ is decreasing. Therefore we get, in a similar way, that

$$\int_0^x f(t)\,dt = 0, \ \forall x \in [0, \infty)$$

and as $f$ is decreasing we have

$$\int_0^x f(t)\,dt \geq x \cdot f(x), \ \forall x \in [0, \infty),$$

from which we obtain $x \cdot f(x) \leq 0$, $\forall x \in [0, \infty)$, a relation that holds for $x > 0$ only if $f(x) = 0$. And so we obtain a new result.

**Problem 23a.** *Let $f : [0, \infty) \to [0, \infty)$ be a decreasing function, having the property that there exists $a \in (0, 1)$ so that (1) is true. Then $f(x) = 0$, $\forall x \in (0, \infty)$.*

**Remark.** Unlike the case of $f$ being increasing, here we do not have $f(0) = 0$ in general. However, Problem 23a is still interesting.

It seems that Solution 2, which allowed the discovery of Problem 23a, is more general than Solution 1. In the second part of this section we will present another solution method for problems of this type, but now let us present the original source problem of this section.

**Problem 23b.** *Let* $a, b \in (0, \infty)$ *with* $a+b < 1$ *and* $f : [0, \infty) \to [0, \infty)$ *an increasing function so that*

$$\int_0^x f(t)\, dt = \int_0^{ax} f(t)\, dt + \int_0^{bx} f(t)\, dt, \ \forall x \in (0, \infty). \tag{4}$$

*Prove that* $f(x) = 0$, *for any* $x \in (0, \infty)$.

**Solution.** Using the arguments from the Solution 2 of Problem 23, by (4) we have

$$F(x) = F(ax) + F(bx), \ \forall x \in (0, \infty) \tag{5}$$

and taking $t_0 < x$ arbitrary, the inequality

$$f(t) - af(at) - bf(bt) \geq (1 - a - b) f(t)$$

gives

$$\int_0^x [\, f(t) - af(at) - bf(bt)]dt \quad \begin{aligned} &\geq (1-a-b)\int_0^x f(t)dt \\ &\geq (1-a-b)\int_{t_0}^x f(t)\,dt \geq (1-a-b)(x-t_0)\cdot f(t_0), \end{aligned}$$

so $f(t_0) = 0$, because $1 - a - b > 0$, $x - t_0 > 0$ and so on. □

We suggest that readers tackle the following conjecture.

**Problem 23c.** *If* $F : [0, \infty) \to [0, \infty)$ *is a continuous function which satisfies* (5) *with* $a + b < 1$, *then is it true that*

$$F(x) = 0, \ \forall x \in (0, \infty)?$$

**Remark.** It is possible that the answer is negative, as the function $F(x)$ in Problem 23b is also increasing — an assumption that we did not ask for in Problem 23c.

## 23.2 The Integral Mean at a Point

Let $f : [0, \infty) \to \mathbb{R}$ be a locally integrable function, i.e., a Riemann integrable function on $[0, x]$, for any $x > 0$. **The integral mean of the function** $f$ on the interval $[a, b] \subset [0, \infty)$ is, by definition (see [4], p. 330),

$$\mu[f\,] = \frac{1}{b-a}\int_a^b f(x)\,dx,$$

while the function $\mu_c[f]: (0,\infty) \to \mathbb{R}$, given by

$$\mu_c[f](x) = \frac{1}{x}\int_0^x f(t)dt, \quad 0 < x < \infty,$$

is called the **Césaro integral mean** of the function $f$ (see [3]). The concept of integral mean appears in the mean value theorem for the Riemann integral: if $f$ is continuous, then there is a point $c \in [a,b]$, so that $\mu[f] = f(c)$.

The Césaro integral mean is an integral transformation which keeps the monotonicity for certain classes of functions (see [2], for instance): if $f: [0,\infty) \to \mathbb{R}$ is continuous and monotonic and $\mu_c[f]: (0,\infty) \to \mathbb{R}$ is the Césaro integral mean of $f$, then $\mu_c[f]$ is also monotonic and has the same type of monotonicity as $f$. The property stays true also when $f$ is not continuous.

We will consider here the limit value of the integral mean, a notion which will be useful for solving Problem 23b.

**Definition.** Let $I \subset \mathbb{R}$ be an interval and $f: I \to \mathbb{R}$ be a locally integrable function. The function $M_f: I \to \overline{\mathbb{R}}$ ( $\overline{\mathbb{R}} = \mathbb{R} \cup \{-\infty, \infty\}$),

$$M_f(x_0) = \lim_{\substack{x \to x_0 \\ x > x_0}} \frac{1}{x - x_0}\int_{x_0}^x f(t)dt\,, \quad \forall x_0 \in I,$$

provided that the limit exists, is called **the integral mean (to the right) of f at** $x_0$.

It is clear that the integral mean of $f$ at $x_0$ is the limit of the integral mean of $f$ on the interval $[x_0, x]$, a limit evaluated when $x$ goes to $x_0$, with values greater than $x_0$.

**Remarks.** 1) The notion of **integral mean at a point** is a little bit inappropriate. It is actually "the derivative of the primitive" of the function $f$.

2) Because the function $f$ is locally integrable, the function

$$F: I \to \mathbb{R}, F(x) = \int_{x_0}^x f(t)\,dt, \quad x \in I$$

is continuous on $I$ and so

$$M_f(x_0) = \lim_{\substack{x \to x_0 \\ x > x_0}} \frac{F(x) - F(x_0)}{x - x_0} = \lim_{h \to 0} \frac{1}{h}\int_{x_0}^{x_0+h} f(t)\,dt.$$

**Theorem 1.** *Let $I \subset \mathbb{R}$ be an interval, $f: I \to \mathbb{R}$ a locally integrable function and $\overline{f}: \mathbb{R} \to \mathbb{R}$ defined by*

$$\overline{f}(x) = \begin{cases} f(x), & \text{for } x \in I, \\ 0, & \text{for } x \neq I. \end{cases}$$

*Then $f$ has primitives over $I$ if and only if*

$$M_{\overline{f}}(x) = f(x), \ \forall x \in I.$$

**Proof.** "$\Rightarrow$": Let $F$ be a primitive of $f$ on $I$. Then $F' = f$ and for all $x \in I$ with $[x - \eta, x + \eta] \subset I$ we have, by the Leibniz–Newton formula, that

$$\frac{1}{h} \int\limits_{x}^{x+h} f(x)dx = \frac{1}{h} \left[F(x+h) - F(x)\right], \ \forall 0 < |h| < \eta$$

and so

$$\lim_{h \to 0} \frac{1}{h} \int\limits_{x}^{x+h} f(x)dx = \lim_{h \to 0} \frac{1}{h} \left[F(x+h) - F(x)\right] = F'(x) = f(x).$$

If $x_0 \in I$ and $x_0 = \inf I$ ($x_0 = \sup I$, respectively) we proceed similarly.

"$\Leftarrow$": Let $a \in I$ be fixed and $F(x) = \int\limits_{a}^{x} f(t)\,dt$, $x \in I$. Then

$$\frac{F(x+h) - F(x)}{h} = \frac{1}{h} \left[ \int\limits_{a}^{x+h} f(t)\,dt - \int\limits_{a}^{x} f(t)\,dt \right] = \frac{1}{h} \int\limits_{x}^{x+h} f(t)\,dt,$$

$\forall x, \, x+h \in I$ and $M_{\overline{f}}(x) = f(x)$. So

$$\lim_{h \to 0} \frac{F(x+h) - F(x)}{h} = M_f(x) = f(x), \ \forall \, x \in I,$$

i.e., $F$ is differentiable at $x$ and $F'(x) = f(x)$. Therefore $F$ is a primitive of $f$ on $I$. □

**Corollary 1.** *If $f : I \to \mathbb{R}$ is locally integrable and has primitives, then*

$$M_f(x) = f(x), \ \forall \, x \in I.$$

**Theorem 2.** *Let $f : I \to \mathbb{R}$ be a monotonic function and $x_0 \in I$. Then*

*a) if $f$ is increasing, then*

$$f(x_0) \le M_f(x_0) \le f(x_0 + 0);$$

*b) if $f$ is decreasing, then*

$$f(x_0) \ge M_f(x_0) \ge f(x_0 + 0).$$

**Proof.** We will tackle only part a). For part b) the proof is similar.

Because $f$ is increasing, it is locally integrable. For any $x \in [x_0, x]$ we have

$$f(x_0) \leq f(t) \leq f(x)$$

and so

$$f(x_0) \cdot (x - x_0) \leq \int_{x_0}^{x} f(t)dt \leq f(x) \cdot (x - x_0), \quad x > x_0,$$

i.e.,

$$f(x_0) \leq \frac{1}{(x - x_0)} \int_{x_0}^{x} f(t)dt \leq f(x),$$

and passing to the limit we get the required inequality. □

**Corollary 2.** *If $f$ is continuous (to the right) at $x_0$, then*

$$M_f(x_0) = f(x_0).$$

**Proof.** Letting $F(x) = \int_a^x f(t)\,dt$, $x > x_0$, $F$ is differentiable to the right at $x_0$ and

$$M_f(x_0) = \lim_{\substack{h \to 0 \\ h > 0}} \frac{F(x_0 + h) - F(x_0)}{h} = F'_d(x_0) = f(x_0).$$

**Remark.** In the proof of Theorem 2 we should have also proved that, in case $f$ is monotonic over $I$, then the integral mean exists at any point $x_0 \in I$.

According to **Corollary 2,** we have to show that this property holds for $x_0 \in I$, even when $f$ is not continuous at $x_0$. The following proposition holds.

**Proposition 1.** *If $f : I \to \mathbb{R}$ is monotonic, then for all $x_0 \in I$ there exists $M_f(x_0)$.*

**Proof.** Because $f$ is monotonic, it is locally integrable, so $f$ is integrable on $[x_0, x]$, for any $x > x_0$, $x \in I$. According to the Lebesgue criterion for the Riemann integral, $f$ is a.e. continuous on $[x_0, x]$.

Let $x_0$ be a point of discontinuity for $f$. As $f$ is monotonic, $f$ has only discontinuities of the first kind, i.e., there exist (finite) lateral limits at $x_0$. We consider

$$\overline{f}(x) : [x_0, x] \to \mathbb{R}, \; \overline{f}(x) = \begin{cases} f(x), & \text{for } x \in (x_0, x], \\ f(x_0 + 0), & \text{for } x = x_0\,. \end{cases}$$

Then

$$\int_{x_0}^{x} f(t)dt = \int_{x_0}^{x} \overline{f}(t)dt$$

and so $M_f(x_0) = M_{\overline{f}}(x_0)$, supposing this last one exists.

But $\overline{f}$ is continuous to the right at $x_0$ and so, according to Corollary 2, $M_{\overline{f}}(x_0)$ exists and is equal to $f(x_0+0)$. □

**Proposition 2.** *a) If $f \geq 0$, then $M_f \geq 0$.*
*b) If $f$ is positive and monotonic, then $M_f$ is also monotonic, of the same monotonicity as $f$.*

**Proof.** a) It is obvious. b) Let $f$ be an increasing function and $x_1 < x_2$. We choose $h > 0$ so that $x_1 + h \leq x_2$. For $t \in [x_1, x_1+h]$, $u \in [x_2, x_2+h]$ we have

$$f(t) \leq f(x_1+h) \leq f(x_2) \leq f(u) \leq f(x_2+h)$$

and so

$$\int_{x_1}^{x_1+h} f(t)dt \leq h \cdot f(x_2) \leq \int_{x_2}^{x_2+h} f(u)du,$$

i.e.,

$$\frac{1}{h}\int_{x_1}^{x_1+h} f(t)dt \leq \frac{1}{h}\int_{x_2}^{x_2+h} f(u)du,$$

from which we get $M_f(x_1) \leq M_f(x_2)$. □

We can now solve Problem 23b using the concept of integral mean (at a point).

**Solution 3.** Let $x_0 > 0$. From (1) we deduce that

$$\int_0^{x_0} f(t)dt = \int_0^{ax_0} f(t)dt + \int_0^{bx_0} f(t)dt$$

which subtracted from (1) gives

$$\int_{x_0}^{x} f(t)dt = \int_{ax_0}^{ax} f(t)dt + \int_{bx_0}^{bx} f(t)dt, \ \forall x > x_0,$$

i.e.,

$$\frac{1}{x-x_0}\int_{x_0}^{x} f(t)dt = \frac{a}{ax-ax_0}\int_{ax_0}^{ax} f(t)dt + \frac{b}{bx-bx_0}\int_{bx_0}^{bx} f(t)dt,$$

from which (as $f$ is monotonic)

$$M_f(x_0) = aM_f(ax_0) + bM_f(bx_0). \tag{6}$$

Because $M_f$ is increasing (as $f$ is increasing), and $a < 1$, $b < 1$, we have $ax_0 \leq x_0$, $bx_0 \leq x_0$, so

$$M_f(ax_0) \leq M_f(x_0) \text{ and } M_f(bx_0) \leq M_f(x_0).$$

Hence from (6),

$$M_f(x_0) \le (a+b)M_f(x_0) \;\Leftrightarrow (1-a-b)\,M_f(x_0) \le 0$$

and as $a+b<1$, i.e., $1-a-b>0$, it results that $M_f(x_0)\le 0$. But $f$ is positive and so $M_f(x_0)\ge 0$, which implies $M_f(x_0)=0$. Using Theorem 2 a) we have

$$0 \le f(x_0) \le M_f(x_0) = 0$$

which implies $f(x_0)=0$. As $x_0$ was taken arbitrary, we deduce that $f(x_0)=0$, $\forall x\in[0,\infty)$. □

In the same way we can solve Problem 23a and the following one.

**Problem 23d.** *Let* $a_1,a_2,\dots,a_n\in(0,\infty)$, *with* $a_1,a_2,\dots,a_n<1$ *and* $\sum\limits_{i=1}^{n} a_i = 1$. *If* $f:[0,\infty)\to[0,\infty)$ *is an increasing function so that*

$$\int_0^x f(t)dt = \sum_{i=1}^{n}\int_0^{a_i x} f(t)\,dt, \quad \forall\, x\in[0,\infty),$$

*then* $f\equiv 0$.

**Remarks.** 1) The notion of integral mean at a point is also inappropriate because $f$, being locally integrable, is Riemann integrable on any interval $[0,a]$ and so, according to the Lebesgue criterion for the Riemann integral, $f$ is a.e. continuous.

This means that, except for a null set $A$, we have $f$ continuous on $[0,a]$ and so $M_f(x)$ coincides at those points with $f(x)$.

Actually, for $x\in[0,a]\backslash A$, $M_f(x)$, is a derivative (of the function $F(x)=\int\limits_0^x f(t)dt$, which is a primitive of $f$ on any interval on which $f$ is continuous).

2) The integral mean of a function at a point, $M_f(x_0)$, is linked to the Césaro integral mean $F(x)$ by the relation

$$M_f(x_0) = \lim_{\substack{x\to x_0\\ x>x_0}} \frac{xF(x)-x_0F(x_0)}{x-x_0},$$

i.e., $M_f$ is the derivative of the function $xF(x)$, where $F$ is the Césaro integral mean.

3) According to Theorem 1, if $f$ is locally integrable and has primitives over $I$, then $M_f(x)=f(x)$, $\forall x\in I$.

If we regard $M_f$ as an operator defined over the set of locally integrable functions, the results of this paragraph show that the set of fixed points of $M_f$ includes the class of primitivable functions.

4) We can also define the integral mean at a point by a symmetrical limit

$$M_f(x_0) = \lim_{h\to 0}\frac{1}{2h}\int_{x_0-h}^{x_0+h} f(t)dt,$$

or by the limit to the left, thus obtaining slightly different concepts.

We will now indicate a few other problems that can be solved in a similar way to the previous ones.

**Problem 23e.** *Find the increasing functions* $f : [0,\infty) \to [0,\infty)$ *for which*

$$\int_0^{2x} f(t)dt = \int_0^{x} f(t)dt, \ \forall\, x \in [0,\infty).$$

**Problem 23f.** *Find the decreasing functions* $f : [0,\infty) \to [0,\infty)$ *for which*

$$\int_0^{2x} f(t)dt = \int_0^{x} f(t)dt, \ \forall\, x \in [0,\infty).$$

**Problem 23g.** *Let* $k \in \mathbb{N}$, $k \geq 2$ *(fixed). Find the decreasing functions* $f : [0,\infty) \to [0,\infty)$ *for which*

$$\int_0^{x^k} f(t)dt = 0, \ \forall\, x > 0.$$

**Solution.** We take $F(x) = \int_0^x f(t)dt, \forall\, x \in (0,\infty)$ and so

$$F(x^k) = F(x), \ \forall\, x \in (0,\infty)$$

from which we get

$$F(x) = F\left(x^{\frac{1}{k}}\right), \ F\left(x^{\frac{1}{k}}\right) = F\left(x^{\frac{1}{k^2}}\right), \ \ldots \ F\left(x^{\frac{1}{k^{n-1}}}\right) = F\left(x^{\frac{1}{k^n}}\right),$$

hence

$$F(x) = F\left(x^{\frac{1}{k^n}}\right)$$

which leads to $F(x) = \lim\limits_{n\to\infty} F\left(x^{\frac{1}{k^n}}\right) = F(1), \ \forall\, x > 0$. Then

$$M_f(x_0) = \lim_{\substack{x \to x_0 \\ x > x_0}} \frac{F(x) - F(x_0)}{x - x_0} = 0, \ \forall \ x_0 \in (0,\infty).$$

As $f$ is positive and increasing, $0 \leq f(x_0) \leq M_f(x_0) = 0$ implies that $f(x_0) = 0$, $\forall\, x_0 \in (0,\infty)$. □

**Problem 23h.** *Find the increasing functions* $f : [0,\infty) \to [0,\infty)$ *having the property*

$$\int_0^{2^k} f(t)dt = \int_0^{3^k} f(t)dt, \ \forall \ x_0 \in (0,\infty).$$

**Solution.** With $F(x) = \int_0^x f(t)dt$, the given relation leads to

$$F(2^x) = F(3^x),\ \forall x_0 \in (0, \infty),$$

where $F$ is a continuous function. Taking $2^x = y \Rightarrow x = \log_2 y$ and $3^x = y^\alpha$, with $\alpha = \log_2 3 \in (1, 2)$, the functional relation becomes

$$F(y) = F(y^\alpha)$$

from which we get

$$F(y) = F\left(y^{\frac{1}{\alpha^n}}\right),\ n \in \mathbb{N}.$$

This finally gives

$$F(y) = F(1),\ \forall\, y \in (0, \infty)$$

and so

$$M_f\,(x) = 0,\ \forall\, x \in (0, \infty)$$

from which we obtain $f(x) = 0$, $x \in (0, \infty)$ and then $f(0) = 0$. The only function which verifies the conditions of the problem is therefore $f \equiv 0$. □

Other exploratory directions would be obtained by considering some functional equations of the type approached in Theme 15. The functional equations that matter are those whose solution is a constant function, a case in which $M_f \equiv 0$, which together with the monotonicity and the sign of $f$ will lead to the conclusion $f \equiv 0$.

## 23.3 Bibliographical Comments

Problem 23 is a simplified form of Problem 23b. The latter was proposed at the $49^{th}$ Mathematics Olympiad, Vaslui, March 1998, by M. Piticari.

The idea of introducing the notion of "integral mean at a point" occurred during grading at the National Mathematics Olympiad in Vaslui, where we also sketched the solution described here.

As far as we know, the notion of integral mean at a point introduced in 23.2 has not been considered effectively until now. Anyhow, some results regarding it can be found in [1] and [4], where it is not defined explicitly. Moreover, Theorem 1 is taken from [4], vol. I, Proposition 9.4.15, together with its proof.

## 23.4 References

1. **V. Arsinte,** *Elementary problems of Integral Calculus.* (Romanian), Editura Universitatii Bucuresti, 1995

2. **M. Becheanu et al.,** *The National Mathematics Olympiad, Forms VII to XII. Questions, Solutions, Marking Schemes.* (Romanian), Editura Paralela 45, 1998

3. **I. Munteanu,** *On the Césaro Integral Mean.* (Romanian), Gazeta Matematica, **85** (1980), no. 12, 483–484

4. **G. Siretchi,** *Integral and Differential Calculus*, vol. 1-2. (Romanian), Editura Stiintifica si Enciclopedica, 1985

5. **S. Lang,***Analysis I.* Addison-Wesley Publ. Co., Reading, Massachusetts, 1968

6. **S. Lang,***Analysis II.* Addison-Wesley Publ. Co., Reading, Massachusetts, 1968

# Theme 24

# Difference and Differential Equations

Recurrent sequences have formed perhaps the most attractive topic of elementary mathematics journals in Romania in the last 3–4 decades (accordingly, not less than four themes in the present book are dedicated to this topic). In spite of the fact that the basic principles of this topic are simple, but not always known by everybody, its variations can still give students or young mathematicians the feeling of discovering new facts.

Actually, this is an advantage of the Romanian mathematics education system, which assiduously cultivates problems with recurrent sequences, linear as well as nonlinear. In other countries this topic is studied in a more detailed manner only at university level, in the frame of the courses Discrete Mathematics or Difference Equations.

The present section — the last of this book — will have a rather different structure in exposing the material and will tend, more clearly than the other sections, to higher mathematics topics, as it uses some differential equations directly related to the recurrence relations considered.

Even if in the Romanian literature the term *recurrent sequence* is widely used, we will prefer here the term *difference equation* to designate this topic.

## 24.1 Difference Equations of First Order

**Problem 24.1.** *Solve the difference equation*

$$x_{n+1} = x_n + a_n,\ n \geq 0, \tag{1}$$

$(a_n)$ *being a given sequence.*

**Solution.** We take $n := 0, 1, 2, \ldots, n-1$ in (1), sum these $n$ equalities side by side, cancel similar terms, to get

$$x_n = x_0 + \sum_{k=0}^{n-1} a_k, \tag{2}$$

which is the solution of the equation (1). □

**Remarks.** 1) The initial problem associated to (1), which asks for those sequences that satisfy $x_0 = b$, has a unique solution, as any first order linear (difference or differential) equation:

$$x_n = b + \sum_{k=0}^{n-1} a_k.$$

2) Equation (1) defines a **generalized arithmetic progression**. If $a_n = r$ (const), $n = 0, 1, 2, \ldots$, then (1) defines the arithmetic progression

$$x_n = x_0 + n \cdot r \quad \text{or} \quad x_n = x_1 + (n-1) \cdot r, \ n \geq 1.$$

3) The continuous analogue of equation (1) is the first order differential equation

$$y' = a(x) \tag{1'}$$

with the solution

$$y(x) = y(x_0) + \int_{x_0}^{x} a(s)ds \tag{2'}$$

which is the continuous analogue of (2). In the case $a(s) = c$ (const) the solution will be

$$y(x) = y(x_0) = (x - x_0) \cdot c.$$

**Problem 24.1a.** *Solve the difference equation*

$$x_{n+1} = ax_n + b_n, \ n \geq 0, \tag{3}$$

*where $a \in \mathbb{R}^*$ is a constant and $(b_n)$ is a given sequence.*

**Solution.** We divide (3) by $a^{n+1}$ and obtain

$$\frac{x_{n+1}}{a^{n+1}} = \frac{x_n}{a^n} + \frac{b_n}{a^n}, \ n \geq 0$$

which is a difference equation of the form (1) and hence

$$\frac{x_n}{a^n} = \frac{x_0}{a^0} + \sum_{k=0}^{n-1} \frac{b_k}{a^k},$$

that leads to

$$x_n = a^n \left( x_0 + \sum_{k=0}^{n-1} \frac{b_k}{a^k} \right), \ n \geq 1. \quad □ \tag{4}$$

**Remarks.** 1) For $a = 1$, from (4) we obtain exactly (2).

2) The continuous analogue of (3) is the differential equation

$$y' - ay = b(x), \quad a = \text{constant}, \tag{3'}$$

whose general solution is

$$y(x) = e^{ax}\left[y(x_0) + \int_{x_0}^{x} b(s)e^{-as}ds\right] \tag{4'}$$

which preserves the form of the difference equation solution (4), replacing the sign $\sum$ by $\int$.

3) The method of solving equation (3) can be applied in some particular cases even if $a$ is not necessarily constant, as shown by the next problem.

**Problem 24.1b.** *Solve the difference equation*

$$x_{n+1} = nx_n + b_n,\ n \geq 0, \tag{5}$$

*where* $(b_n)$ *is a given sequence.*

**Solution.** We divide (5) side by side by $n!$ to obtain

$$\frac{x_{n+1}}{n!} = \frac{x_n}{(n-1)!} + \frac{b_n}{n!},\quad n \geq 1$$

which is a difference equation of the form (1), and so the solution is

$$x_n = n!\left[x_0 + \sum_{k=1}^{n} \frac{b_{k-1}}{k!}\right],\quad n \geq 1.$$

□

**Problem 24.1c.** *Solve the difference equation*

$$x_{n+1} = a_n x_n + b_n,\ n \geq 0, \tag{6}$$

*where* $(a_n)$ *and* $(b_n)$ *are given sequences with* $a_n \neq 0$, $n \geq 0$.

**Solution.** We use Proposition 2 in [2] which asserts that, if $(a_n)$ is a sequence of nonzero real numbers, then there exists a unique sequence $(c_n)$ such that $c_p = 1$ and

$$\frac{c_n}{c_{n+1}} = a_n,\ \text{for any } n \geq p$$

(where $p \in \mathbb{N}$ is fixed). Indeed, as $a_n \neq 0$, it results that $c_n \neq 0$ and so

$$\frac{c_p}{c_{p+1}} \cdot \frac{c_{p+1}}{c_{p+2}} \cdots \frac{c_{n-1}}{c_n} = a_p a_{p+1} \cdots a_{n-1},$$

which gives

$$c_n = \left(a_p a_{p+1} \cdots a_{n-1}\right)^{-1},\ n \geq p+1.$$

Using $(c_n)$ previously determined, relation (6) can be written as

$$c_{n+1}x_{n+1} = c_n x_n + b_n c_{n+1}, \ \ n \geq 0$$

which has the form (1). After all computations (here $p = 0$) we get

$$x_n = \frac{1}{c_n}\left(x_0 + \sum_{k=1}^{n-1} b_k c_{k+1}\right),$$

that is

$$x_n = \left(\prod_{k=0}^{n-1} a_k\right) \cdot x_0 + \sum_{k=1}^{n-1} b_k \cdot \left(\prod_{i=k+1}^{n-1} a_i\right), \ \ n \geq 1. \qquad \square \quad (7)$$

**Remarks.** 1) The continuous analogue of (6) is the non-homogenous first order linear differential equation

$$y' = P(x)y + Q(x) \tag{6'}$$

whose general solution is (see for example [9])

$$y(x) = y(x_0)e^{\int_{x_0}^{x} P(s)ds} + \int_{x_0}^{x} \left(Q(t)e^{\int_{t}^{x} P(s)ds}\right) dt,$$

which preserves, by analogy, the form of the solution (7) of the correspondent discrete equation.

2) The method previously presented for scalar equations can be extended to vector difference equations of the first order. For the sake of simplicity, we present it below in the two dimensional case.

**Problem 24.1d.** Let $(A_n)$, $(B_n)$ *be two sequences of matrices,* $A_n \in \mathcal{M}_2(\mathbb{R})$, $B_n \in \mathcal{M}_{2,1}(\mathbb{R})$, $n = 0, 1, 2, \ldots$, *such that* $A_n$ *is invertible and* $A_n A_m = A_m A_n$ *for all* $n, m \in \mathbb{N}$. *Solve the difference equation*

$$X_{n+1} = A_n X_n + B_n, \ \ n \geq 0,$$

*where* $(X_n)$ *is a sequence in* $\mathbb{R}^2$.

**Solution.** First of all, we search for a sequence of matrices $(C_n)$ satisfying $C_n = I_2$ (identity matrix) and

$$C_n \cdot C_{n+1}^{-1} = A_n, \ \ n \geq 0. \tag{8}$$

Since $A_n$ is invertible, from (8) we obtain

$$\left(C_0 C_1^{-1}\right)\left(C_1 C_2^{-1}\right) \ldots \left(C_{n-1} C_n^{-1}\right) = A_0 A_1 \ldots A_n$$

which gives, due to the associativity property,

$$C_n = (A_0 A_1 \ldots A_n)^{-1}, \ \ n \geq 1. \tag{9}$$

Since $A_n$, $A_m$ commute, we also have

$$C_n \cdot C_{n+1}^{-1} = C_{n+1}^{-1} C_n, \ n \geq 0. \tag{10}$$

Indeed, in view of (9) relation (10) is equivalent to

$$(A_0 A_1 \dots A_n)^{-1} \cdot A_0 A_1 \dots A_n A_{n-1} = A_0 A_1 \dots A_n A_{n-1} \cdot (A_0 A_1 \dots A_n)^{-1}$$

which, based on the fact that

$$(A_0 A_1 \dots A_n)^{-1} = A_n^{-1} A_{n-1}^{-1} \dots A_0^{-1},$$

becomes

$$A_n^{-1} A_{n-1}^{-1} \dots A_0^{-1} \cdot A_0 A_1 \dots A_n A_{n-1} = A_0 A_1 \dots A_n A_{n-1} \cdot A_n^{-1} A_{n-1}^{-1} \dots A_0^{-1},$$

and, using the commutativity of matrices $A_n$ , this implies

$$A_{n-1} = (A_0 A_1 \dots A_n \cdot A_{n-1}^{-1} \dots A_0^{-1}) A_{n-1},$$

which is obvious.

Now using (8) and (10) the equation in Problem 24.1d becomes

$$X_{n+1} = C_{n+1}^{-1} C_n X_n + B_n$$

and hence, by multiplying it side by side by $C_{n+1}$, we get

$$C_{n+1} X_{n+1} = C_n X_n + C_{n+1} B_n$$

which is a vector equation of the form (1). Therefore

$$C_n X_n = X_0 + \sum_{k=0}^{n-1} C_{k+1} B_k,$$

that is

$$X_n = (A_0 A_1 \dots A_n) \left( X_0 + \sum_{k=0}^{n-1} (A_0 A_1 \dots A_k)^{-1} B_k \right), \ n \geq 1,$$

which is an expression of the form (4). □

Based on the previous general result we can solve problems of the following type:

**Problem 24.1e.** *Find sequences* $(x_n)$, $(y_n)$ *which satisfy the recurrence relations*

$$x_{n+1} = x_n \cos \frac{n\pi}{4} + y_n \sin \frac{n\pi}{4} + 1,$$

$$y_{n+1} = -x_n \sin \frac{n\pi}{4} + y_n \sin \frac{n\pi}{4} + n, \ n \geq 0,$$

*where* $x_0, y_0$ *are given.*

**Solution.** We take

$$X_n = \begin{pmatrix} x_n \\ y_n \end{pmatrix}, \quad A_n = \begin{pmatrix} \cos\frac{n\pi}{4} & \sin\frac{n\pi}{4} \\ -\sin\frac{n\pi}{4} & \cos\frac{n\pi}{4} \end{pmatrix}, \quad B_n = \begin{pmatrix} 1 \\ n \end{pmatrix}, \quad n \geq 0$$

and the system of recurrence relations is equivalent to

$$X_{n+1} = A_n X_n + B_n$$

which fulfils all assumptions in Problem 24.1d. Since

$$A_n = A^n, \text{ where } A = \begin{pmatrix} \cos\frac{\pi}{4} & \sin\frac{\pi}{4} \\ -\sin\frac{\pi}{4} & \cos\frac{\pi}{4} \end{pmatrix},$$

we shall have

$$A_0 A_1 \ldots A_k = A^{1+2+\ldots+k} = A^{\frac{k(k+1)}{2}},$$

hence

$$\begin{pmatrix} x_n \\ y_n \end{pmatrix} = \begin{pmatrix} \cos\alpha_n & \sin\alpha_n \\ -\sin\alpha_n & \cos\alpha_n \end{pmatrix} \cdot \begin{pmatrix} x_0 \\ y_0 \end{pmatrix} + \sum_{k=0}^{n-1} \begin{pmatrix} \cos\beta_{k,n} & \sin\beta_{k,n} \\ -\sin\beta_{k,n} & \cos\beta_{k,n} \end{pmatrix}, \quad (11)$$

where

$$\alpha_n = \frac{n(n+1)\pi}{8}, \quad \beta_{k,n} = \left[\frac{n(n-1)}{2} - \frac{k(k+1)}{2}\right]\frac{\pi}{4}, \quad n \geq 1, \ k \geq 0.$$

As the sums in (11) can be fully evaluated, we can find the sequences $(x_n), (y_n)$ in a simple form. □

## 24.2 Equations of Order Greater than 1

**Problem 24.2a.** *If $(x_n)$ is a sequence of real numbers such that there exist two numbers $\alpha$ and $\beta$ for which*

$$x_n = \alpha \cdot x_{n-1} + \beta \cdot x_{n-2}, \ \forall n = 2, 3, \ldots,$$

*then*

$$x_n = \frac{a^n - b^n}{a - b} x_1 - ab\frac{a^{n-1} - b^{n-1}}{a - b} x_0, \ \forall n = 1, 2, \ldots,$$

*where $a \neq b$ are the roots of the equation $x^2 = \alpha x + \beta$ (the so-called characteristic equation associated to the difference equation).*

*If $a = b$, then $x_n = na^{n-1}x_1 - (n-1)a^n x_0$, $n = 1, 2, \ldots$.*

**Solution.** If $a, b$ are the roots of the characteristic equation

$$x^2 - \alpha x + \beta = 0,$$

then $\alpha = a + b$, and $\beta = -ab$, such that the recurrence relation can be written as

$$x_n - a x_{n-1} = b(x_{n-1} - a x_{n-2}), \ \ n = 2, 3, \dots .$$

If we take $y_n = x_n - a x_{n-1}$, then we have $y_n = b y_{n-1}$ which is a homogenous relation of the form (3). Hence

$$y_n = b^{n-1} y_1$$

which gives $x_n - a x_{n-1} = b^{n-1} y_1$, $n = 1, 2, \dots$. This is again an equation of the form (3), this time an inhomogeneous one. By (4) we get

$$x_n = a^n \left( x_0 + \sum_{k=0}^{n-1} \frac{b^{k-1} y_1}{a^k} \right),$$

from which we obtain exactly the expression in the statement, by computing the sum in the right-hand side for $a \neq b$ and $a = b$, respectively. □

**Remarks.** 1) If $a$ and $b$ are complex numbers (conjugates of each other, since $\alpha$ and $\beta$ are real numbers), then by writing them in the trigonometric form

$$a = \bar{b} = r(\cos t + i \sin t)$$

we obtain, by the de Moivre formula, that

$$x_n = r^{n-1} \frac{\sin nt}{\sin t} \cdot x_1 - r^n \frac{\sin(n-1)t}{\sin t} \cdot x_0, \ \ n = 1, 2, \dots .$$

2) The method used in solving Problem 24.2a can be applied to solving difference equations with non-constant coefficients as well.

**Theorem 1.** *Let*

$$x_{n+2} - a_n x_{n+1} + b_n x_n = c_n, \ \ n \geq 0 \tag{12}$$

*be a difference equation with* $(a_n), (b_n)$ *satisfying*

$$a_n^2 - 4b_n \geq 0, \ \ \forall\, n \geq 1 \tag{13}$$

*and at least one of the roots of the characteristic equation* $r^2 - a_n r + b_n = 0$ *being independent of* $n$. *Then the general solution of* (12) *is given by*

$$x_n = r_1^n \left( x_0 + \sum_{k=0}^{n-1} \frac{v_k}{r_1^{k+1}} \right), \ \ n \geq 0, \tag{14}$$

*where*

$$v_n = \prod_{k=0}^{n-1} r_2(k) \left( u_1 - r_1 u_0 + \sum_{k=0}^{n-1} \frac{c_k}{\prod_{i=1}^{k} r_2(i)} \right) \tag{15}$$

*and* $r_1$ *and* $r_2(n)$ *are the roots of the characteristic equation.*

**Solution.** Under the given hypotheses let $r_1$ and $r_2(n)$ be the two real roots of the characteristic equation, $r_1$ being independent of $n$. As

$$a_n = r_1 + r_2(n), \quad b_n = r_1 r_2(n),$$

by substituting in (12), we get

$$x_{n+2} - r_1 x_{n+1} - r_2(n)(x_{n+1} - r_1 x_n) = c_n.$$

We denote now

$$v_n = u_{n+1} - r_1 u_n \tag{16}$$

and thus obtain that $v_n$ is the solution of the difference equation

$$v_{n+1} - r_2(n) v_n = c_n,$$

which has the form (3). Using (4), we obtain exactly (15), since $v_0 = u_1 - r_1 u_0$. With $v_n$ already computed, we go back to the first order equation (16)

$$u_{n+1} - r_1 u_n = v_n$$

which is again of the form (3), and by (4) we will get exactly (14). □

As an application of Theorem 1, let us solve

**Problem 24.2b.** *Solve the difference equation*

$$(2n+1)\cdot x_{n+2} - 4(n+1)\cdot x_{n+1} + (2n+3)\cdot x_n = (2n+1)(2n+3), \quad n \in \mathbb{N}.$$

**Solution.** The equation can be written in the form

$$x_{n+2} - \frac{4(n+1)}{2n+1} x_{n+1} + \frac{2n+3}{2n+1} x_n = 2n+3$$

and, by denoting

$$a_n = \frac{4(n+1)}{2n+1}, \quad b_n = \frac{2n+3}{2n+1}, \quad c_n = 2n+3,$$

we have

$$a_n^2 - 4b_n = 1 > 0, \quad \forall\, n \in \mathbb{N}.$$

The roots of the characteristic equation are

$$r_1 = 1 \text{ and } r_2(n) = \frac{2n+3}{2n+1}, \text{ respectively.}$$

Therefore, all the conditions in Theorem 1 are satisfied, hence using (15) we get

$$v_n = \prod_{k=0}^{n-1} \frac{2k+3}{2k+1} \left[ v_0 + \sum_{k=0}^{n-1} \frac{2k+3}{\prod_{i=0}^{k} \frac{2i+3}{2i+1}} \right],$$

that is

$$v_n = (2n+1)(v_0 + n), \ n \geq 0,$$

and if we substitute it in (14) we get the general solution

$$x_n = x_0 + \sum_{k=0}^{n-1}(2k+1)(v_0+k),$$

and after all computations

$$x_n = (1-n^2)x_0 + n^2 \cdot x_1 + \frac{n(n-1)(4n+1)}{6}, \ n \geq 1. \qquad \square$$

**Problem 24.2c.** *Find the general solution of the difference equation*

$$\begin{aligned}\left[(n+1)^3 - n^3\right] x_{n+2} - \left[(n+2)^3 - n^3\right] x_{n+1} + \left[(n+2)^3 - (n+1)^3\right] x_n \\ = \left[(n+1)^3 - n^3\right] \cdot \left[(n+2)^3 - (n+1)^3\right], \qquad n \in \mathbb{N}.\end{aligned}$$

**Solution.** By denoting

$$a_n = \frac{6n^2+12n+8}{3n^2+3n+2}, \ b_n = \frac{3n^2+9n+7}{3n^2+3n+1}, \ c_n = 3n^2+9n+7, \ n \in \mathbb{N},$$

we can see that the equation has the form (12). Condition (13) is satisfied, as

$$a_n^2 - 4b_n = \left(\frac{3n+3}{3n^2+3n+1}\right)^2 > 0, \ \forall\, n \in \mathbb{N},$$

but none of the roots of the characteristic equation

$$r_1(n) = 1 + \frac{9n+7}{6n^2+6n+2} \ \text{ and } \ r_2(n) = 1 + \frac{3n+3}{6n^2+6n+2}$$

is constant. Therefore, Theorem 1 cannot be applied in this case. However, using Green's formula [1], we can find the general solution as

$$x_n = \alpha + \beta n^3 + \frac{1}{9}n^2(n-1)(3n+1), \ n \in \mathbb{N}. \qquad \square$$

**Remark.** By exploiting the same idea as in solving Problem 24.2b, we can solve the continuous analogous case as well.

**Problem 24.2d.** *Solve on* $\mathbb{R}\backslash\left\{-\frac{1}{2}\right\}$ *the differential equation*

$$(2x+1)y'' - 4(x+1)y' + (2x+3)y = (2x+1)(2x+3).$$

**Solution.** The equation can be written as

$$y'' - \frac{4(x+1)}{2x+1}y' + \frac{2x+3}{2x+1}y = 2x+3,$$

that is

$$y'' - y' + \frac{2x+3}{2x+1}(y' - y) = 2x + 3.$$

We denote $z = y' - y$ and so the previous equation is reduced to the linear (non-homogenous) first order differential equation

$$z' + \frac{2x+3}{2x+1}z = 2x + 3,$$

whose solution is given (see, for example, [9], p. 17) by

$$z(x) = Ce^{-\int_0^x P(s)ds} + e^{-\int_0^x P(s)ds} \cdot \int_0^x Q(t)e^{\int_0^x P(s)ds} dt,$$

where $P(x) = \frac{2x+3}{2x+1}$ and $Q(x) = 2x + 3$.

Afterwards, $z(x)$ being determined in this way, we have to solve with respect to the unknown $y$ the first order linear differential equation

$$y' - y = z(x),$$

in order to obtain the general solution of the initial equation. □

**Remark.** If we carefully read the solutions of Problems 24.2b and 24.2c we can notice that the assumption "at least one of the roots of the characteristic equation is constant" is not necessary for solving the difference equations there.

Indeed, if we can find $r_1(n)$ and $r_2(n)$ such that

$$\begin{cases} a_n = r_1(n) + r_2(n), \\ b_n = r_1(n-1) \cdot r_2(n), \end{cases} \tag{17}$$

then the difference equation can be written as

$$x_{n+2} - r_1(n)x_{n+1} - r_2(n)[x_{n+1} - r_1(n-1)x_n] = c_n$$

and, by denoting

$$v_n = x_{n+1} - r_1(n-1)x_n,$$

it is reduced to an inhomogeneous first order linear difference equation

$$v_{n+1} - r_2(n)v_n = c_n.$$

Knowing $v_n$, we have to solve with respect to $x_n$ the equation

$$x_{n+1} - r_1(n-1)x_n = v_n$$

which is of the same type.

We shall illustrate this technique by means of the next problem.

**Problem 24.2f.** *Solve the difference equation*

$$x_{n+2} - a_n x_{n+1} + b_n x_n = c_n, \quad n \geq 0,$$

*with $a_n$ and $c_n$ as in Problem 24.2c and*

$$b_n = \frac{3n^2 + 9n - 5}{3n^2 + 3n + 1}.$$

**Solution.** It is easy to check that (17) holds for

$$r_1(n) = \frac{3n^2 + 15n + 7}{3n^2 + 3n + 1}, \quad r_2(n) = \frac{3n^2 + 9n - 5}{3n^2 + 3n + 1}$$

and hence we can solve the difference equation (the computations are left for the reader). □

**Remarks.** 1) If $(a_n), (b_n)$ are given, in order to find $r_1(n)$ and $r_2(n)$ that satisfy (17) it is necessary to solve the nonlinear difference equation

$$r_1(n) = a_n - \frac{b_n}{r_1(n-1)}, \quad n \geq 1,$$

which is a more difficult task.

2) The same method as the one presented before may be applied to solve difference equations of order $q \geq 2$ as well as for their continuous analogues.

At the end we invite our readers to adapt all the results in this section for the $q$-order difference equations, see also [2].

## 24.3 Bibliographical Comments

Problem 24.1b was presented by V. Pop from the Technical University of Cluj-Napoca at the Symposium "Modern Aspects of Teaching Mathematics", Satu Mare, 28 May 1984. His idea was then extended to more general problems; for a survey of some related results see [2].

Problem 24.2a is Theorem 2 in [4], p. 63, in an slightly different form. Even the solution method is taken from [4].

Theorem 1 is proved in a more general setting in our paper [3], which is the source of the most results and applications included in Section 24.2.

Problem 24.2b is taken from [1], p. 72, Example 2.10.2, where it is solved by means of the Green's formula for difference equations.

The idea of proving Theorem 1 is similar to the method used in [7] by the authors in order to find the general solution of a second linear homogenous differential equation with constant coefficients.

## 24.4 References


1. **R.P. Agarwal,** *Difference Equations and Inequalities. Theory, Methods and Applications*, Marcel Dekker, 1992

2. **V. Berinde,** *On a Class of Nonlinear Recurrences.* (Romanian), Lucr. Semin. Creativ. Mat., **4** (1994–1995), 89–94

3. **V. Berinde,** *A Method for Solving Second Order Difference Equations*, in New Developments in Difference Equations and Applications (S.S. Cheng, G. Ladas & S. Elaydi, eds.), Gordon & Breach Publishers, 1998, pp. 41–48

4. **D. Busneag, I. Maftei,** *Themes for Mathematical Competitions.* (Romanian), Editura Scrisul Romanesc, 1983

5. **M. Chirita, D. Popovici,** *On the Systems of Recurrent Sequences.* (Romanian), Gazeta Matematica (Serie metodica), **15 (94)** (1996), no. 2, 82–90

6. **A. Markusevici,** *Recurrent Sequences.* (Romanian), Biblioteca SSM, 1953

7. **F. Popovici, M. Bencze,** *On the Second Order Homogenous Linear Ordinary Differential Equations with Constant Coefficients.* (Romanian), Gazeta Matematica (Seria metodica), **15 (94)** (1997), no. 3, 198–201

8. **N. Rosca,** *Recurrent sequences.* (Romanian), Gazeta Matematica, 80 (1975), no. 6, 208–213

9. **I.A. Rus, P. Pavel,** *Differential Equations.* (Romanian), Editura Didactica si Pedagogica, Bucuresti, 1982

10. **V. Sadoveanu,** *Homogenous Recurrent Sequences.* (Romanian), Gazeta Matematica, **83** (1978), no. 5, 187–190

# Addendum

## Basic Principles Regarding Creativity in Solving Problems

Apart from the necessary mathematical rigor and accuracy, solving a problem should respect at least the following three principles:

1. The algorithmicity principle;
2. The generality principle;
3. The generalization principle.

The first principle, that of **algorithmicity**, implies that the solution should be clearly structured, highlighting the successive stages, so that the basic lines constitute a well-traced sketch for solving the problem.

As solving any problem means performing a succession of operations, algorithmicity means exactly grouping and processionally arranging them.

Respecting this principle leads to a clear solution, easy to follow and apply to other problems of the same type, by its author as well as by others.

The second principle, that of **generality**, implies that once we have solved a problem and the solution has a coherent method, we should investigate to what extent that method can be applied to problems related to the initial one. Therefore, it is necessary that we search for and gather as many related problems as possible, to validate the generality of the method.

If our solution has a singular character, we will not be able to apply it to all the related problems, but if it has general arguments, deductions and implications, we can build a model method, applicable to all the problems of the family. This, of course, by making certain corrections, changes and improvements.

Collecting problems of the same type around a model method is the basis of writing any textbook or exercise book and it is also, in our opinion, one of the most used and most useful method of training students for mathematics contests. Having such a collection and an illustrated solution method for a couple of problems, the student is left with the duty to approach the other problems, starting off with the basic idea of the method. The efforts in applying this principle when solving a problem provide the first creative feature of the solution.

The third principle, that of **generalization**, even though related to that of generality, cannot be identified with it all the time. This results clearly from the examples given in all themes of this book.

This principle says that once we have solved the problem, we should not consider the solution complete before at least the following steps are made. These are

- analyzing all the components of the problem;
- examining and highlighting the necessary and/or sufficient character of each argument we used;
- examining and highlighting the extent to which the data given in the text was used at different stages of the solution.

This allows, among other things:

- weakening some conditions from the text;
- exchanging some conditions with other equivalent ones;
- getting rid of some useless conditions (the ones that are not used effectively in the solution, but occur in the text due to the way in which the author himself solved the problem);
- identifying the arguments, deductions, implications that stay valid in a more general case;
- linking the conclusion, or some partial results, to other results in order to get some connections and interconnections which broaden the horizon of the problem.

To weaken the conditions means, generally, to find the sufficient conditions and to replace them with weaker ones, but still sufficient, or necessary and sufficient. By making this detailed analysis and synthesizing the results, we will obtain new problems, already solved, whose assumptions are weaker than those of the initial one, but whose results are stronger.

But the most important gain is that in this manner we can get suggestions for new problems, of the same type or simply related, which we only have to formulate and, especially, to solve.

Such an approach to a mathematics problem is called in the spirit of this book a *creative approach.* This leads to new results, being actually the basis of mathematical research. All the results in this domain are obtained, consciously or not, by applying the generality and mostly the generalization principle.

We must say that we cannot apply these last two principles to just any problem. This means that not any problem can be the starting point for obtaining new problems.

But a sure thing is that all those who propose problems, the authors of notes and articles in mathematics, do apply, more or less, these principles. The reader can decide whether the ideas presented in this book are valid or not by taking a close look at *Gazeta Matematica* (the youth series, and especially the methodical series). We strongly recommend also the articles published in the magazine *Lucrarile Seminarului de Creativitate Matematica,* as they too illustrate these ideas.

Of course, one can object that, when applying the generality principle to adjust the method and make it applicable to a larger class of problems, we actually take exactly the steps involved in the generalization principle. Therefore, even if one could consider that the two are exactly the same thing, we will keep this delimitation, illustrated in a couple of examples. Essentially, what holds our reasoning is that generality can be considered an attribute of the solution method, while generalization refers to the text of the problem and to the notions and facts involved: the *generality* of the method means that we can use it when solving other related problems, but the *generalization* does not take us out of the domain of the problem.

## Conclusions

We have followed these principles, more or less, in all the 24 sections of this book, showing how a creative approach to elementary mathematics problems should look. Any researcher — in the field of mathematics or not — follows these principles in his work, consciously or not. We recommend that readers read the book [8] whose basic ideas are in spirit the same as those exposed here. But, first of all, the interested reader should see G. Polya's excellent book [4] where he exposes in detail a heuristics of solving mathematics problems.

The necessity of solving a problem may well appear in any domain of human activity and, regardless of the nature of this problem, one has to involve creativity in order to get a solution. So most of the ideas presented in this book can be well applied in a non-mathematical context, too.

## References

1. **H. Banea,** *About Problems of Didactical Mathematics.* (Romanian), Gazeta Matematica (Seria metodica), **1** (1980), no. 3, 99–104

2. **H. Banea,** *About Exhaustive Problems.* (Romanian), Gazeta Matematica (Seria metodica), **6** (1985), no. 1–2, 31–34

3. **A. Netea,** *Problems from Gazeta Matematica - Starting Point in Scientific Research.* (Romanian), **4** (1983), no. 3–4, 127–130

4. **G. Polya,** *Mathematical Discovery. On Understanding, Learning, and Teaching Problem Solving.,* John Wiley & Sons, 1965

5. **G. Polya,** *How to Solve it. A New Mathematical Method.* Reprint of the second edition, Princeton University Press, 1988

6. **E. Rusu,** *The Psychology of Mathematical Activity. (*Romanian), Editura Stiintifica, Bucuresti, 1969

7. **E. Rusu,** *How to Think and Solve* 200 *Problems. (*Romanian), Editura Albatros, 1972

8. **F.W. Stevenson,** *Exploratory Problems in Mathematics,* National Council of Teachers of Mathematics (USA), 1992

9. **N. Teodorescu,** *Textbooks of Problems and Solving Methodologies. (*Romanian*),* Gazeta Matematica, **5** (1984), no. 3–4, 99–105

10. **A. Zugravu,** *Between Habit and Creation. (*Romanian*),* Gazeta Matematica (Seria metodica), **2** (1981), no. 2–3, 56–60

11. **V. Berinde,** *Basic Principles Regarding Solving of Problems. (*Romanian*),* Lucr. Semin. Creativ. Mat, **3** (1993–1994), 89–98

GPSR Compliance

*The European Union's (EU) General Product Safety Regulation (GPSR) is a set of rules that requires consumer products to be safe and our obligations to ensure this.*

*If you have any concerns about our products, you can contact us on ProductSafety@springernature.com*

In case Publisher is established outside the EU, the EU authorized representative is:

Springer Nature Customer Service Center GmbH
Europaplatz 3
69115 Heidelberg, Germany

**Batch number: 10726298**

Printed by Printforce, the Netherlands